THE MECHANICS OF
ENGINEERING SOILS

THE MECHANICS OF ENGINEERING SOILS

P. LEONARD CAPPER

T.D., M.Sc. (Eng.), F.I.Struct.E., M.I.C.E.

Formerly Senior Lecturer, University College, University of London

and

W. FISHER CASSIE

C.B.E., LL.D., Ph.D., F.R.S.E., F.I.C.E., F.I.Struct.E., P.P.H.E.

Emeritus Professor of Civil Engineering, University of Newcastle upon Tyne

SIXTH EDITION

LONDON

E. & F. N. SPON LTD

A Halsted Press Book
John Wiley & Sons, Inc., New York

First published 1949
Second edition 1953
Reprinted three times
Third edition 1960
Reprinted twice
Fourth edition 1963
Reprinted once
Fifth edition 1969
Reprinted three times
Sixth edition 1976

E. & F. N. Spon Ltd
11 New Fetter Lane, London EC4P 4EE

© *1976 P. Leonard Capper and W. Fisher Cassie*

Printed in Great Britain
at the
University Printing House, Cambridge
(Euan Phillips, University Printer)

ISBN 419 10990 0

Distributed in the U.S.A. by Halsted Press,
a Division of John Wiley & Sons, Inc., New York

Library of Congress Cataloging in Publication Data
Capper, Percival Leonard.
　The mechanics of engineering soils.

　Bibliography: p.
　Includes index.
　1. Soil mechanics. I. Cassie, W. Fisher, joint author. II. Title.
TA710.C26　1976　　624'.151　　76-10238
ISBN)-470-15109-9

Contents

v

Contents

Preface to the Sixth (S.I.) Edition

Since the First Edition in 1949, this book has passed through five editions with additional printings. Changes in the text have been made from time to time to maintain the text in touch with current developments. The change to S.I. units in the Sixth Edition gives the opportunity for a more radical revision, appropriate now in view of the many strands of development the subject has shown in recent years. The early chapters on fundamental principles change but little, but those in the latter portions of the book deal with applications to practice. They are concerned with the correlation of data and particularly with judgement, assessment and prediction as related to the behaviour of soils on particular sites. In these areas of study there have been changes in instrumentation, in knowledge and in techniques of decision-making. The text emphasizes the importance of obtaining relevant and reliable data during site investigation, the first step in the assessment of any building site.

As in all previous editions, this text is aimed at the first or second year student at University or Polytechnic. It presents the range of the subject in sufficient substance and detail to allow of first attempts at translating data from the site into decisions in the solution of constructional problems. Esoteric studies and complex methods which may be employed only in unusual conditions are not much stressed. These will be encountered later when the student has made himself familiar with the basic principles and applications of soil mechanics, and has achieved that sense of confidence which it is the aim of this book to encourage.

More than a score of problems, carefully selected, lead from study to application. For further problems the reader is directed to the authors' companion volume, *Problems in Engineering Soils,* which is also in an S.I. edition. Particular thanks goes to Mr C. O. Vallally of Northumbrian Drilling Contractors for his useful discussions and his assistance in preparing the text.

It is with deep regret that I report the death of the senior author,

Mr P. L. Capper. He was active in the preparation of this Sixth Edition, contributing ideas and text to the revisions. Mr Capper was a pioneer in the teaching of soil mechanics, having given one of the first courses of university lectures (at University College, London) in 1938. His early work in the subject was acknowledged at the 25th Anniversary of the British Geotechnical Society.

1976   W. Fisher Cassie

Notation

A Pore-pressure coefficient
Flow nets: cross-sectional area carrying flow
Area of Pile

A_1 Area of foundation

A_p Settlement: area of applied pressure diagram

A_s Piles: circumferential area of cylindrical shaft

a Sheet piling: depth of anchor bar below top of sheet piling.
Circle analysis:
distance from centre of slip to cohesive force.
Spacing of resistivity electrodes

B Breadth or diameter of footing or pile

$B, \bar{B}$ Pore-pressure coefficients

b Slips: breadth of slice

C Cohesion: force

C_w Cohesion: force between soil and retaining wall

C_c Consolidation: compression index

C_e Consolidation: expansion index

C_p Penetration tests: static cone penetration pressure

c Cohesion: apparent cohesive stress

c' Cohesion: cohesive stress referred to effective stress conditions

c_m Cohesion: mobilized cohesive stress

c_u Cohesion: cohesive stress; undrained test

c_v Coefficient of consolidation

D Diameter
Slips: factor defining position of critical slip circle

d Sheet piling: depth of penetration
Slips: distance of centroid from centre of rotation
Piling: diameter of pile
Consolidation: length of drainage path

E Earth pressure: passive resistance of wedge
Slips: inter-slice force

e Void ratio

F Factor of safety

f Coefficient of friction

G_s Specific gravity of particles

H Slips: height of bank
Flow: head

h Slips: height of slice
Retaining wall: vertical height
Flow: head
Soil: thickness of sample of layer of soil

I Current in geophysical tests

i Hydraulic gradient
Slope of bank

K Influence factor

K_a Earth pressure: coefficient of active pressure

K_o Earth pressure: coefficient of pressure at rest

K_p Earth pressure: coefficient of passive earth pressure

K_ϕ $\tan^2(45° - \phi/2)$

k coefficient of permeability

L Foundations: length of rectangular footing
Slips: length of slip surface

l Piles: effective length
Slips: length of slip surface of slice

M Flow: width of seepage channel

m_v Coefficient of compressibility

N Seepage flow: loss of head
Slips: normal component of weight of slice
Standard Penetration number

N' Slips: Effective force as normal component of weight of slide

N_c, N_q, N_γ Bearing capacity factors

n Porosity
Slips: Factor defining position of critical slip circle

P Concentrated point load
Resultant force on slip surface
Weight of pile

P_w Thrust due to water pressure

P_a Earth pressure: active thrust on wall (force)

P_p Earth pressure: force of passive resistance (force)

p Consolidation pressure: overburden pressure

p_a Earth pressure: active pressure on the wall

p_p Earth pressure: passive resistance (pressure)

pF Soil suction: logarithm of suction head

Q Flow: total discharge

q Discharge in unit time
Intensity of contact pressure
Bearing capacity

R Slips, earth pressure: resultant reaction on plane of rupture or slip surface
Radius of slip surface

r Radius

R, R_s, R_p Pile resistance

r_u Pore-pressure ratio

r_p Intensity of pressure under pile

S Soil suction
Degree of sensitivity

s Degree of saturation
Shearing resistance (stress)
seconds

s_t Settlement after time t

T Slips: tangential component of weight of slice (force)

T_v Consolidation: time factor

t Time interval

U_v Consolidation: degree of consolidation

u Pore-pressure

V Volume
Voltage in geophysical tests

V_a Volume of air

V_d Volume of dried soil sample

V_s Volume of solids

v Flow: velocity

JP fordresher

W Earth pressure: weight of wedge or disturbed volume

Piles: load to be supported

w Moisture content

x, y, z Rectangular co-ordinates

z Depth in soil sample or stratum

z_c Slips: depth of tension cracks

α Fraction of total pressure

Earth pressure: angle between active thrust and the vertical

Slips: inclination of slip surface of slice to horizontal

β Earth pressure, slips: angle of surface to horizontal or angle of surcharge

δ Earth pressure: angle of friction between soil and back of wall, or between soil and shaft of pile

Elastic compression of pile

ν Poisson's ratio

ϵ Strain

γ Bulk weight density

γ_d Dry weight density

γ_w Weight density of water

η Viscosity of liquid

ϕ Flow: potential kh

ϕ Angle of shearing resistance

ϕ_u in undrained conditions

ϕ_{cu} in consolidated-undrained conditions

ϕ_d in drained conditions

ϕ' referred to effective stress

ϕ_r residual shearing resistance

ρ Bulk mass density

ρ_d Dry mass density

ρ_w Mass density of water

σ Total normal pressure

σ' Effective pressure or stress

σ_1 Vertical pressure: maximum principal stress

σ_3 Lateral pressure: minimum principal stress

σ_2 Intermediate principal stress

θ Retaining walls: angle between back of wall and horizontal

Slips: angle subtended by arc of slip

CHAPTER 1

The scope of soil mechanics

Soil Mechanics may be defined as the application of mechanics, hydraulics and geology to engineering problems relating to soils. The term 'soil' is used here in its normal civil engineering sense, that is, to describe the sediments and deposits of solid particles produced by the disintegration of rock.

The deposits whose behaviour is outlined in this volume are extremely variable in composition, and it was this discouraging heterogeneity which so long retarded the scientific study of sands, gravels, and clays. Until a clear conception of the nature of these materials was developed in the form of principles and conclusions of wide application, no scientific study of the soil as a load-carrying structure was possible. Collection and correlation of data were slow, but the gradually increasing scale of man's constructional activities forced him to apply to large works the experience he had gained on smaller constructions. This step was not always successful, and investigation of failures of retaining walls, foundations, and embankments gradually led to the accumulation of sufficient data regarding the nature of the soil and its behaviour to allow of the growth of the science of soil mechanics.

1.1 THE NATURE AND PROPERTIES OF SOILS

It is now well established that the various approaches to the study of soils and their behaviour must be concurrent and interrelated. There can be no true understanding of the behaviour of soils without a clear

1

appreciation of the nature and variability of the materials themselves. The primary study, then, must be that of classification.

It was found that broad terms such as *gravel, sand, silt,* or *clay* conveyed to the minds of engineers varying conceptions of the nature of the material in question. Classification by means of the pre-dominant particle-size of the soil is therefore a fundamental step towards the elimination of ambiguities. The further recording of the grading or distribution of all particle-sizes throughout the sample gives an even sounder basis of classification. When the soil is of the clay type (having a large proportion of colloidal particles) its appearance and nature at varying moisture contents give further information by which it can be described and its probable behaviour predicted.

Exact and adequate identification and classification can be made only if satisfactory samples of the soil are available. Boring methods, types of tools for extraction of satisfactory samples, and methods used for the determination of properties of the soil *in situ* have all been subject to critical study.

Having determined the classification of the material under investi-gation, the engineer then ascertains its engineering properties, the most important of which are *shear strength, moisture/density relation-ship, permeability,* and *compressibility*. These may sometimes be quoted as secondary aspects of classification, but their significance lies rather in their influence on the behaviour of the soil under engineering conditions. From the results of such tests the probable behaviour of the soil when subjected to known loading conditions may be estimated. In the tests for these properties the soil sample must be as undisturbed as possible, a condition which has to be taken into account in the development of sampling techniques.

Methods for determining the properties of soil *in situ* have been developed partly in order to avoid the errors introduced by the process of sampling, and partly to obtain the values of properties not easily obtained from sampling.

1.2 PROBLEMS IN SOIL MECHANICS

Having studied the nature of the soil with which he must deal, the engineer turns to the problems which the science of soil mechanics is called upon to solve. These concern the soil in its natural state and position, or in the state imposed upon it by the engineer. The problems may be studied in two groups:

Problems of equilibrium or stability and
Problems of elastic and plastic deformation.

Problems of stability

Here we are concerned with the complete and sudden collapse of the resisting capacity of the soil. Failure of stability is accompanied by large and sudden movements. A retaining wall moves forward; the footing of a heavily loaded column sinks into the ground, which heaves up around it: the sloping side of an embankment slumps in a characteristic slip surface In attempting to prevent failure of stability the engineer must be able to answer three questions:

What is the load imposed on the soil?
What magnitude (and distribution) of stress is induced in the soil by this load?
What resistance can the soil afford?

The first two of these questions have been answerable for many decades. The answer to the first depends on the nature of the structure to be erected and the weight of the soil itself, both readily defined. The answer to the second depends on the mathematical analysis of stress within an ideal material. While it is recognized that stresses so determined may not agree closely with the stresses found in practice, thorough investigation of stability failurers in recent years has done much to check the data obtained by the mathematician. The third question is answered more and more completely by the study of the nature and behaviour of soils in different parts of the world, both in the laboratory and on constructional work. The engineer can give in advance definite assurance of the stability of certain types of structure, and his range of action is continually widening. The search for more rapid methods of estimating properties of soil *in situ* goes on.

Problems of deformation

Soils act both as elastic and as plastic materials, and the changes in volume which they undergo in both these states when under load result in cumulative deformations of a magnitude large enough to be important to the engineer. Deformation problems involve a study of moisture content and moisture movement in the soil, and much attention has been paid to this aspect. The consolidation of clays – one of the most important effects of moisture movements – is sufficiently understood for both the magnitude and rate of settlement of engineering structures to be predicted if sufficient data are available. The nature of the deformation of soils also has an important effect on the distribution of pressure on footings and retaining walls, and such distribution in time affects the further settlements and movements of the engineering structure concerned. The point of application of the resultant pressure on a retaining wall is, for example, often more important than its magnitude.

1.3 LINK WITH LOCAL KNOWLEDGE AND EXPERIENCE

The young and inexperienced engineer may, in his enthusiasm for the 'scientific' work described in this book, fall into error. There is no suggestion that modern soil mechanics methods will inevitably supersede all the older and more empirical procedures. Civil engineering is still more of an art than a science, and is likely to remain so, especially when the engineer deals with large-scale geological formations which are not easily amenable to mathematical treatment. Modern methods of soil mechanics can give considerable assistance in the solution of problems of an unfamiliar scale, or in unfamiliar country, but local knowledge of soil conditions, amassed over the centuries is of first importance, to be supplemented rather than displaced by the findings of soil mechanics. Further, the variability of soil over an area and the changing conditions as the depth below the surface increases present a picture far removed from the 'homogeneous, isotropic, elastic, and semi-infinite mass' of the mathematician. In these circumstances again, care is required; the link with local knowledge must be strong; facile conclusions from a few tests may well be misleading unless they are interpreted in the light of experience.

1.4 HISTORICAL BACKGROUND

The properties and behaviour of soils have exercised the minds of builders from time immemorial. An early allusion to the erosion of foundations is found in the parable of the man who built his house upon the sand, with disastrous consequences!

Very little attempt appears to have been made to deal with soil problems on a scientific basis until towards the end of the eighteenth century, when Coulomb published his theory of earth pressure and suggested his well-known law of the failure of soils. Rankine's theory followed about a century later, and these two classical theories still form the basis underlying modern methods of estimating earth pressure.

Coulomb, Rankine and Bell

The classical theories of Coulomb and Rankine envisaged cohesionless soil such as dry sand. Numerous alternative theories and graphical methods have since been devised, many of them merely adaptations of the original theories with some of the assumptions modified. A notable advance was made in 1914 by Bell, who took cohesion into account. If due attention is paid to the limits of their applicability the theories of Coulomb, Rankine and Bell are still considered sound, and corrections are easily made when required to allow for deviations from the ideal assumptions.

In the theories of Coulomb and Rankine it was assumed that the surface of rupture of the soil is plane. While this is reasonably correct for cohesionless soil, it is well known that cohesive soils tend to slip along a curved surface. Swedish engineers began to study this about 1911, and Petterson proposed the assumption of a circular arc as the trace of the surface of slip. On this assumption practical methods, both analytical and graphic, have been devised and many slip failures have been investigated.

Terzaghi

Even up to the First World War, most soil problems, other than earth pressure on retaining walls, were solved by empirical methods, usually based on inspection of the soil on the site, sometimes assisted by rather primitive loading tests. Scientific investigation of soils had been mainly confined to the geological, chemical, and physical aspects in relation to agricultural requirements. During the last half century, however, much progress has been made in the scientific study of soils and in the development of systematic testing. The application of theory and experimental data to practical engineering design has also considerably developed.

It was Karl Terzaghi who first brought theory and empiricism together. His publication in 1925 of *Erdbaumechanik* marked the definite beginning of a new understanding of this aspect of engineering geology. Terzaghi's influence on English-speaking engineers was marked by such authoritative papers as *The Science of Foundations – Its Present and Future* given to the American Society of Civil Engineers, and by his research and consulting work. Terzaghi, with justification, earned the title of 'The Father of Soil Mechanics'.

Until the development of soil mechanics, little scientific data were available regarding the stability of foundations. Rankine's theory for the minimum depth of foundations was of little practical value, for it was based on the assumption of a dry, granular soil. The designer of foundations had to rely very largely on experience gained from buildings in the same locality. When local experience was not available he depended upon loading tests. Such tests frequently gave misleading results because of the scale effect, and sometimes because of the existence of unsuspected weaker strata at some depth below the foundation.

Study of the shearing resistance of soils and of the mechanism of failure when foundations are overloaded, together with the development of the technique of site investigation and testing, have led to more reliable methods of design.

An important stage in the application of soil mechanics to foundation engineering was the publication by Terzaghi in 1925 of his theory of consolidation. This theory enables estimates to be made of the probable amount and rate of settlement of buildings founded upon clay soils.

During the last century great advances have been made in the design and construction of pavements and road surfaces, but until recently little attention was paid to the underlying natural material. In present day practice the principles of compaction and stabilization are applied to both pavement and subgrade. The results of research work on soils and the improved mechanical equipment now available have enabled these processes to be carried out more extensively and in a more efficient manner.

Research and experience

Research work on subgrade materials was started about 1920 by the then U.S. Bureau of Public Roads. The tests devised by the Swedish physicist Atterberg in 1911 (liquid limit, plastic limit, etc.) and other simple tests were correlated with the properties of natural soils in relation to road engineering.

In England, the Road Research Board was formed in 1933 under the Department of Scientific and Industrial Research. Work was carried out first at the Building Research Station and since that time also at the Transport and Road Research Laboratory.

During the Second World War, much valuable experience was gained from the construction of airfields in various parts of the world. Shortage of labour, transport difficulties, the necessity for rapid completion, and often interference from the enemy, introduced many new problems. Much research work has been carried out on the improvement of the subgrade, the design and construction of pavements, the efficiency of mechanical equipment, and the development of soil tests. An interesting feature is the use of mobile laboratories for carrying out the necessary soil tests on the site. Several of these were in use during the later stages of the Second World War.

The civil engineer is now more dependent on the numerical results of tests than were his predecessors. These tests often amplify and illuminate the findings of experience or link new problems with those already solved. One of the outstanding difficulties encountered in the making of tests is that of extracting a sample of soil in such a way that it is truly representative of the material in question. The difficulty of doing this effectively has encouraged a trend towards testing *in situ* instead of in a laboratory.

1.5 SINCE THE THIRTIES

A conference on soil mechanics was held at Harvard University, Cambridge, Mass., in 1936, at which papers by engineers and scientists from all parts of the world were read and discussed. This conference did much to stimulate interest in the subject and may be regarded as a landmark in its development. A second International Conference, at which the progress of twelve years was reported, took place at Rotterdam in 1948, and similar conferences have since been held every four or five years.

In addition to organizing international conferences, the International Society of Soil Mechanics and Foundation Engineering, through its branches in various countries, co-ordinates research and disseminates information. The British section, the British Geotechnical Society, produces a journal, *Geotechnique,* in which articles and papers on research and on the practical applications of soil and rock mechanics are published. In December 1974, the British Geotechnical Society held its 25th anniversary meeting. It was there recalled how the beginnings of the thorough study and teaching of soil mechanics in Great Britain were based on the work of Dr L. F. Cooling and Dr A. W. Skempton in the soil physics section of the Building Research Station (1933) and also on the soils laboratory set up by Dr W. H. Glanville at the Road Research Laboratory. Teaching of soil mechanics at British Universities commenced in the period 1938-40, at University College, London by P. L. Capper, and at the Universities of Sheffield (Professor N. S. Boulton), Glasgow (Dr W. McGregor) and Durham (King's College, Newcastle) (A. H. A. Hogg). The teaching at Glasgow was taken up in 1944 by Dr H. B. Sutherland, and Professor W Fisher Cassie set up a laboratory and teaching at Newcastle in 1943. Teaching by textbooks was mostly from American publications, such as those by Plummer and Dore (1940) and Krynine (1941). The outstanding works were *Theoretical Soil Mechanics* by Karl Terzaghi (1943) and *Soil Mechanics in Engineering Practice* by Terzaghi and Peck (1948). In the U.K., several texts began to appear after the war, with *An Introduction to Soil Mechanics*, Dr W. L. Lowe-Brown (1945), *Practical Problems in Soil Mechanics*, H. R. Reynolds and P. Protopapadakis (1948) and *Soil Mechanics for Civil Engineers*, Dr Bernard H. Knight (1948). *Mechanics of Engineering Soils* was first published in 1949.

The Building Research Establishment at Garston, near Watford, and the Transport and Road Research Laboratory at Crowthorne, Berkshire, are the principal British centres of research on soil problems. The Soils Section of the former carries out investigations on soil mechanics in general and with special reference to earth pressure, stability of slopes and foundation problems. The Soils Section of the Transport and Road Research Laboratory deals with the applications of soil mechanics to roads and airfields.

Engineering schools at universities and polytechnics teach soil mechanics as a regular part of the civil engineering curriculum, and laboratories have been established both for instruction in routine tests and for research. Government departments, municipal authorities, civil engineering contractors, and consulting engineers have equipped their own laboratories for the carrying out of routine tests and for the investigation of special problems.

1.6 WHAT THIS BOOK DOES FOR YOU

Since its first edition, this book has always been directed to providing assistance to the young engineer who comes to soil mechanics for the first time. The chapters take him through enough of the theoretical presentation to show him (or her) that there is a fundamentally accepted basis to the science of the subject. As an academic study, soil mechanics provides as much mental exercise and controlled thinking as does the study of hydraulics or the theory of structures.

Here the similarity ends. In structures, most of the materials are fabricated by man to a specification, and their use is controlled. In hydraulics, the chief material is one which obeys laws amenable to mathematical study. By contrast, such theories as have been devised for soil mechanics must, of necessity, be based on an assumed high measure of uniformity. Soil mechanics in practice deals with a very wide range of highly heterogeneous materials. It is not surprising to find discrepancies existing between theoretical predictions and what happens on the site.

The successful practitioner of soil mechanics is he who admits theory and tests to be his guides, but whose knowledge of site conditions is the mainstay of his advice and proposals. He is, for example, sensitive to the disturbing effect of the presence of water; his first intuitive action is to look at the drainage of the landscape in which his site lies. The absence of an intuitive 'feel' for these forces and others which influence the behaviour of large masses of soil, has led in the past, to many failures.

To use this book effectively, you must bear in mind several aims. The book is intended to encourage you to the accomplishment of these aims.

(1) Develop a viewpoint which is more geological than mathematical. Take the landscape as the unit. Consider the origins, composition, past deformation and future stability of the land forms you see around you. Observation can be assisted by geological publications, but the development of observation is of first importance.

(2) Remember the high place you must give to the study of water in soils. Realize the forces and pressures which static or moving water can produce. Have a mental picture of how water is likely to be

operative, out of sight, in the soil of your landscape. Consider how its power can be controlled in the design of stable structures, and how, by ill-advised action, a destructive force might be released.

(3) Consider well what information you need to allow you to advise on the use of a site, how this information is to be obtained with adequate accuracy and how you are to present the facts in clear and unequivocal terms. All construction starts (or should start) with a Site Investigation Report. It must be written by you or read by you. You must know how to produce it and how to judge the productions of others.

(4) Do not study this book merely to permit you to pass an examination. Make its message part of your mental background so that you are 'soaked' in its conceptions. Study for knowledge and not for examinations.

(5) Master each chapter before passing on to the next. In particular, re-work the Application problems. Then, using slightly different values of the data, find out what happens to the result. Learn to know the influence which one parameter has on another.

(6) The mass of your studies should be directed to the twin aims of

Preserving stability $\left\{\begin{array}{l} \text{in foundations} \\ \text{in soil structure} \end{array}\right.$

Preventing unacceptable deformation $\left\{\begin{array}{l} \text{in foundations} \\ \text{in soil structures} \end{array}\right.$

These are aims which are justified economically, and are of importance. Studies of soils must be aimed at a final product: the knowledge of how to predict their behaviour. When you have reached this stage, you will have done well!

CHAPTER 2

Physical state and classification of soils

It is pointed out in Chapter 1 that the classification of soils according to their distinctive properties is of great value to the engineer. As the properties of a soil are very largely determined by the predominant size of particle and the grading, or proportions of the various sizes, a standard system of classification and nomenclature of particle-sizes is also essential.

The following brief account of the geological origin and composition of soils will be helpful when considering methods of classification and identification.

2.1 GEOLOGICAL ORIGINS

The soils with which the civil engineer is concerned were originally produced by the disintegration of rocky portions of the Earth's crust. Some of the rocks so turned into soil were themselves made from soils by the deposition and consolidation of small particles, and are known by geologists as secondary and tertiary soils. After initial disintegration the products of rock decay may be settled quietly through still water, consolidated by pressure, or dispersed and graded by the action of moving water and ice. The final product is the clay, sand, marl, gravel, or other soil which forms or supports the works of the civil engineer.

On level rock surfaces the disintegrating effect of weathering may produce soils having no tendency to move elsewhere, especially if bound in place by vegetation. Such *residual* soils show a change in character from the solid rock below to the organic soil of the surface, but no minerals foreign to the parent rock are found.

In mountainous, windswept, or stream-washed areas there is considerable movement of the disintegrated particles. This *erosive* movement may deposit the material a long distance from its source of origin, may change its particles from angular to rounded shapes, and may sort and grade the various sizes.

Another cause of disintegration is chemical action. Some rocks, such as limestone, are readily attacked by weak acids; carbonic and humic acids are present in rainwater. Sedimentary rocks of the sandstone type consist of inert particles of silica impervious to weathering, but cemented by material which is liable to be eroded or dissolved.

Physical and chemical weathering and erosion have very different effects according to the prevailing climatic conditions. In arid desert country, for example, there is little or no chemical action, and salts are not washed away by rain or rivers, but are left as crystalline deposits in the sand. On the other hand, in tropical regions, where rainfall is heavy, there is a continuous and cumulative *leaching* out of soluble material, and the bedrock is often weathered to great depths.

Of the various soil-forming agencies, water is probably the most productive. When streams and rivers enter a lake or the sea the load of sediment carried is deposited in graded sizes, the largest particles being deposited first and the smaller particles later, when the velocity of the water has sufficiently decreased. Thus on the geologically younger rivers, or on higher reaches, gravels and coarse sands are found, but on lower or older parts of the course, where the velocity is low, silts and clays predominate. In various parts of the world wide-spreading deltaic deposits occur, where, over centuries, the lighter materials have settled out in marine or estuarine conditions.

Main classes of soil
Certain widespread classes of soil are spoken of by generic terms which cover a wide variety of types, but which indicate sufficiently well one recognizable material. Some of these are:

> *Laterite*: weathered red soil found in tropical countries. Excessive leaching by rain has reduced the soil to an accumulation of sesquioxides.
> *Black-cotton soil*: black – or sometimes brown – soil, which

shows marked volume changes with change in moisture content.

Loess: silty material deposited by wind action. Particle-sizes lie between narrow limits (0·01 to 0·05 mm). The colour is a light brown.

Boulder clay: stony clay showing no distinct stratification and resulting from glacial action. It may be overconsolidated.

Till: glacial deposits consisting of a mixture of all types of soil: gravel, sand, silt, clay, and boulders.

Clay minerals

Differences in properties exist between two cohesive soils of apparently the same physical composition. The need for a secondary physical stsyem of classification – liquid and plastic limits – stems from these differences between apparently similar materials, caused by the composition and the structure of the clay minerals and the forces of attraction between the particles.

Clays are formed of minerals of complex composition. Their composition and nature have been studied by X-ray analysis, and it is clear that they occur as hydrated aluminium silicates, existing in the sheets or flakes. The atomic structure of clays is two-dimensional, and their physical form is thus a plate structure such as in seen in mica.

The plates are formed of alternate layers of simpler compounds, the three important ones being:

Magnesium layer $Mg_3(OH)_6$ – brucite;
Aluminium layer $Al_2(OH)_6$ – gibbsite;
Silicon layer $Si_2O_3(OH)_2$.

According to the arrangement of these layers, clay minerals with different compositions are produced.

There are, broadly, three groups of clay minerals. When silica predominates and there are two silicon layers to one layer of gibbsite (with some brucite) we have the *montmorillonite group*. Montmorillonite, formed by the weathering of ferromagnesian minerals, is well known by its popular name of 'Fullers Earth'. In this group are *montmorillonite, beidellite,* and *nontronite*. These are all similar, and are widespread in clay strata. Bentonite, well known for its use in various modern constructional processes, is composed chiefly of montmorillonite.

The second group comprises the china clays and is known as the

13

kaolin group. *Kaolinite* contains less silica than montmorillonite, having been formed by the weathering of alkaline felspars, and the silicon and aluminium layers alternate. Other minerals slightly differing from kaolinite, but belonging to the same group, are *dickite* and *halloysite.*

The third group is known generically as *illites.* These are less fully understood, but the base-exchange properties of illite are less marked than for montmorillonite. The layers of illite are also more firmly held together, so that they do not break down so readily under remoulding.

2.2 COARSE- AND FINE-GRAINED SOILS

Residual or transported soils, deposited in lake, river, or sea, or formed by wind, ice, and frost, present an infinite variety of particle-size gradings, ranging from boulders to colloidal particles.

As a general basis of classification, soils may be divided into three main classes:

> Coarse-grained or non-cohesive soils;
> Fine-grained or cohesive soils;
> Organic soils, e.g. peat.

Coarse-grained soils are composed of rock fragments varying in size from boulders down to gravel and sand. Quartz, because of its hardness, is the predominant mineral in the composition of many gravels and sands, particularly when the particles are well rounded. Sharp gravels and sands which have not undergone so much wear often contain, in addition, other minerals which enter into the composition of the parent rock. In fine sand deposits the particles are often more angular than in the coarser varieties, since the film of water between the fine grains tends to protect them against abrasion.

Fine-grained soils include silts and clays. *Silt* is the name given to a type of soil (intermediate between fine sand and clay) which exhibits distinctive properties. Its mineral composition is more variable than that of fine sand, since chemical as well as mechanical disintegration often enters into its formation.

Clays are fine-grained soils which possess plasticity, especially when moist. Deposits of clay have been originally deposited in water and then gradually compressed by the weight of the material above them. This process is known as *consolidation*. The soil is said to be *fully* or *partially* consolidated, according to whether or not a state of equilibrium has been reached. In the course of geological history, some of the overlying material may have been eroded. Deposits in which the consolidation pressure has thus been partially relieved are known as *overconsolidated* or *preconsolidated*.

The converse of consolidation (where the bond between particles is strengthened) is the loosening of contact between clay particles. The leaching out of salts from clays deposited originally in marine conditions can weaken the bond which is dependent on the natural electrolyte in the soil. Any disturbance can bring such clays into the *quick* condition where the last remaining strength is lost and the clay flows like a liquid.

From the point of view of structure, clays may exhibit *laminations* or *fissures*. Laminated clay occurs in thin layers, as the name suggests, each thin layer being capable of being peeled from the mass of clay. Fissured clays contain irregularly spaced fissures or cracks, easily seen when a lump is broken. In the natural state fissured clays are usually stiff and of high cohesive strength. The fissures are normally closed, but when opened slightly by excavation, change of loading, or other causes, water enters the cracks, and softening and disintegration set in.

Organic soils. The topsoil nearly always contains organic matter which occurs partly as partially decomposed vegetable matter and partly as humus, a dark amorphous material derived from the decomposition of plant and animal matter. This organic matter is of vital importance to the agriculturist, but usually of less consequence to the engineer, since the topsoil is removed when engineering works are constructed. Frequently, however, deposits of peat are encountered in which the solid matter is nearly all organic. Such soils are extremely treacherous and troublesome from the engineering point of view.

Characteristics of coarse- and fine-grained soils

For the theoretical study of soil mechanics it is often convenient to consider the two primary types, sand and clay, as typical of the coarse-grained and fine-grained classes. The main differences in their physical characteristics may be summarized as follows:

Sand	Clay
Void ratio low	Void ratio high
Negligible cohesion when clean	Marked cohesion, depending on water content
Internal friction high	Internal friction low
Not plastic	Plastic
Only slightly compressible	Very compressible
Compression takes place almost immediately on application of load	Compression takes place slowly over a long period
Permeable to water	Practically impermeable

Field identification of soil types

Gravel, sand, and sand-gravel mixtures are readily identified by inspection. The correct classification of soils containing fine-grained material, however, involves quantitative tests, with qualitative indications as an extra means of identification.

Fine sand may have slight cohesion when damp, or in dry lumps, but such lumps are easily powdered between the fingers; the particles are visible and have a distinctly gritty feel.

Silt is most readily identified by the 'dilatancy' test: shake a pat of moist silt horizontally in the palm of the hand. Moisture will come to the surface, but can be made to recede by pressing the pat with the fingers.

Clay does not exhibit the property of dilatancy; it is smooth and greasy to the touch, and sticks to the fingers when moist.

Organic soils are distinguished by their coarse fibrous texture and dark colour, and sometimes by the distinctive odour of decaying vegetation. This odour is often intensified by heating the soil.

2.3 CLASSIFICATION OF NATURAL SOILS

Several systems of classification of natural soils have been devised to suit the requirements of various types of engineering project.

Broad classification of soil types

Table 2.1 shows the broad classification of soils into the three main classes, their characteristics and principal sub-divisions.

There are of course, in addition, many composite types of soil which are natural mixtures of the primary types.

Table 2.1 *Broad classification of soil types*

Non-cohesive	Gravels	More than 50% of particles coarser than 2 mm
	Coarse sands	Most particles between 2 and 0·5 mm
	Medium sands	Most particles between 0·6 and 0·2 mm
	Fine sands	Most particles between 0·2 and 0·06 mm
Cohesive	Silts	Most particles less than 0·06 mm. Particles invisible or barely visible to the naked eye. Gritty to touch. Exhibit dilatancy. May show slight cohesion when dry.
	Clay	Predominant influence of particles less than 0·002 mm. Smooth to touch. Plastic. No dilatancy. Considerable cohesion when dry.
Organic	Peats	High compressibility. Fibrous. Brown and black colour.

Detailed classification systems

Modern systems of classification of natural soils are based on the original classification proposed by A. Casagrande for the construction of roads and airfields. The primary and composite types are shown in detail, with data for their identification and guides to their properties. A feature of this system is the designation of the soil type by two capital letters. The first indicates the main type, and the second classifies, the sub-division within that type. The symbols are as in Table 2.2.

Table 2.2 *Detailed classification of soil types*

	Main terms	*Qualifying terms*
Coarse-grained soils	Gravel, G	W, well-graded
	Sand, S	P, poorly graded
		M, mixtures
Fine-grained soils	Silt, M*	L, low plasticity
	Clay, C	H, high plasticity
Organic soils	Peat, Pt	

The main term M(*), for silt, derives from the Scandinavian term mo.

Possible combinations of terms, which give indications to the soils engineer as to how the soils will behave are for example:

SW Sandy soil: well graded;
CH Clay: high plasticity;
GP Gravel mixture: poorly graded.

Unified classification system
This comprehensive classification is a modification of the Casagrande system. It is shown in Table 2.3. A further sub-classification is shown in the plasticity chart (Fig. 2.5). There, the fine-grained soils are further sub-divided according to their *liquid limits*.

2.4 DISTRIBUTION OF SOIL PARTICLES

Natural soils are mixtures of particles of various sizes, but we have seen that the properties of a soil are to a very great extent determined by the predominant particle-size in its composition. The experimental determination of the particle-size distribution is therefore an important factor in soil classification. Such analysis is, however, insufficient to give all the information necessary for the detailed classification of a soil, and other simple tests are required, especially if the soil is of a fine-grained type.

In view of the importance of particle-sizes it is necessary to have a system of nomenclature for the various grades or fractions comprising particles lying between certain specified size limits. A number of arbitrary systems have been suggested, differing mainly in the sizes specified as the limits of the constituent grades. Careful distinction must be made between the classification of particle-sizes and the classification of natural soils for engineering purposes as described in the preceding article.

Unfortunately the terms *silt, clay, sand,* and *gravel* are used in two distinct senses:

(i) as the designation of particles whose sizes fall between certain specified limits;
(ii) as descriptions of naturally occurring soils, which are mixtures of grains of different sizes, exhibiting certain definite characteristics associated with those names.

Table 2.3 *Unified soil classification*

Major divisions, groups and typical names	Group symbols	Visual and physical characteristics
1. Coarse-grained soils		
Gravel and gravelly soils		
Well-graded gravels or gravel-sand mixtures, little or no fines	GW	Large particles, easily seen; majority of particles larger than 1·5 mm
Poorly-graded gravels or gravel-sand mixtures, little or no fines	GP	
Silty gravels, gravel-sand-silt mixtures	GM	
Clayey gravels; gravel-sand-clay mixtures	GC	
Sand and sandy soils		
Well-graded sands or gravelly-sands, little or no fines	SW	Most of the particles can be seen without the aid of a magnifier, soils feel gritty to the fingers
Poorly graded sands or gravelly-sands, little or no fines	SP	
Silty sands, sand-silt mixtures	SM	
Clayey sands, sand-clay mixtures	SC	
2. Fine-grained soils		
Silts and clays (LL less than 50)		
Inorganic silts, very fine sands, rock flour, silty or clayey fine sands, or clayey silts with slight plasticity	ML	Not gritty to the fingers, but can be rolled into threads when moist. Shrinkage cracks appear on drying
Inorganic clays of low to medium plasticity, gravelly clays, sandy clays, silty clays, and lean clays	CL	
Organic silts and organic silt clays of low plasticity	OL	
Silts and clays (LL greater than 50)		
Inorganic silts, micaceous or diatomaceous fine sandy or silty soils, elastic silts	MH	Greasy to the touch. Can be rolled easily into threads when moist. Shrinks on drying. More than 40% clay particles
Inorganic clays of high plasticity, fat clays	CH	
Organic clays of medium to high plasticity, organic silts	OH	
3. Highly organic soils		
Peat and other highly organic soils	Pt	Dark and fibrous

It is preferable to use terms such as *silt sizes* and *clay sizes* when the former meaning is intended.

Particle-size classification

The system which has been adopted by the British Standards Institution is one originally put forward by the Massachusetts Institute of Technology (Fig. 2.1). The term *mo,* is a Scandinavian word used to describe a range of particle-size intermediate between fine sand and silt. In German, this material is known as *schluff.* It does not occur in the B.S. classifications.

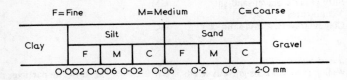

Fig. 2.1 Particle-size classification.

In a paper by Glossop and Skempton, systems of particle-size classification are discussed in relation to the physical properties of the fractions. The authors conclude that the B.S. system is the best suited for engineering purposes, since the boundaries of its main divisions correspond approximately to important changes in the soil properties. It also has the advantage that the particle-sizes defining the various categories increase, at each stage, by a multiple of about three. Thus, logarithmic paper is not required, and ordinary graph paper may be used, the distance between 0·002 and 2 mm being divided into six equal parts, and further similar divisions being used for the larger sizes. In the chapter on laboratory testing, a similar range of sizes is given with a multiple of approximately two. This can also be shown on ordinary graph paper with equal divisions giving equal multipliers. No great accuracy is lost by this device, although the multipliers are not strictly equal throughout.

The analysis for this classification is carried out in two stages:

(i) the separation of the coarser fractions by sieving on a series of standard sieves;
(ii) the determination of the proportions of the finer particles by a sedimentation process, generally known as *fine analysis.*

The methods employed for these separations are discussed in Chapter 14.

Grading curves

The results of the particle-size analyses are plotted as grading curves as shown in Fig. 2.2, the abscissae representing particle-size to a logarithmic scale and the ordinates, percentage by weight finer than the corresponding particle-size.

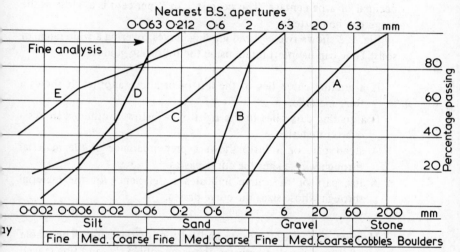

Fig. 2.2. Grading curves.

The positions occupied by the curves on the chart indicate the types of material represented. Those curves which lie higher or farther to the left show gradings of relatively finer material, while those which are lower or towards the right indicate coarser material.

The general slope of the curve is an indication of the grading or range of particle-sizes of which the soil is composed. From an engineering point of view, a *well-graded* mixture is one containing an assortment of particles covering a wide range of sizes. On the grading curve chart, the curve for a well-graded soil stretches at a more-or-less uniform slope across a wide range of sizes. Such a soil usually has greater strength and stability than a *poorly-graded* soil in which the grains are more uniform in size, and the curve is steep.

The shape of a grading curve can present to the engineer a mental picture of the appearance of the material.

Interpretation of grading curves

The aim of a grading analysis is to allow the engineer to obtain a mental picture of the type of soil and to be able to predict its behaviour. This involves drawing a chart, so there is little to be gained by obtaining the proportions of the various sizes of particles to a decimal of a percentage. The nearest 1 or 2 per cent is as close as the graph can be plotted.

Fig. 2.2 shows five stylized grading curves covering a wide range of soils. They can be interpreted, using the following guides:

If a grading curve lies to the left of or above another, it shows a finer material.

If a grading curve lies to the right of or below another, it shows a coarser material.

A steep part of a curve shows a preponderance of the material through whose sites the curve passes.

A flat part of the curve indicates a deficiency of the material through whose sizes the curve passes.

Material A: Look at the top and bottom of the curve. The whole range of this material is between the large cobble size and fine gravel, with a little coarse sand – less than 20 per cent of the weight. In the stockpile this would look very coarse with more or less equal quantities (20 per cent) of coarse and medium gravel, and 30 per cent of fine gravel. This is a well-graded material within the cobble/gravel range; it has nearly equal proportions of all sizes within that range. It could not be called well-graded without qualification as it has no finer material.

Material B: At first glance this can be seen to be poorly graded for the curve is steep. There is a large proportion of coarse sand amounting to 60 per cent of the weight of the sample. A further 18 per cent is coarse gravel, and the rest (22 per cent) is in the medium and fine sand and silts range. In the hand this would appear to be a very coarse sand with some finer material. It would behave as a highly granular material.

Material C: The long range of this curve and the uniform slope shows that this is a well-graded material. About 8 per cent of the weight of the sample is in the fine gravel category, 45 per cent evenly distributed over the three categories of sand, 27 per cent in the three silt divisions, and 20 per cent of clay. In the hand this would be a well-graded sand/silt, with some clay particles – to about a fifth of the total sample. However, the silt and clay proportion (total of 47 per cent) is enough to give this material cohesive properties. When damp and compressed in the hand it would retain its shape. The proportion of silt would be a danger signal for any shallow foundations. Apart from the possibility of long-term settlement, such a material would be sensitive to frost and might show *frost heave* when close below the surface, as, for example in a road foundation.

Material D: This is another cohesive material with 4 per cent clay and 12 per cent fine sand. The rest, 84 per cent, lies in the silt region. The material would look very fine in the hand, and be poorly graded because of the high proportion of silt it contains. Note the steepening of the curve over the silt range. Frost heave and considerable settlement might well be envisaged if foundations were placed in the material. The foundation engineer would, on seeing this curve, consider the possibility of piling through such unacceptable material.

Material E: Half of sample E is of clay, and the rest is spread uniformly in the silt and fine/medium sand regions. This is a typical cohesive soil, perhaps a till, and all the precautions required for clays must be observed. It might be exploited successfully for second grade fill where earth structures are required. The high proportion of silt would prevent its being classed as first class impervious and compactable material for, say, the core of a dam. Sample E would be difficult to handle when wet.

The student of soil mechanics should carry out exercises such as these, assessing soils from their grading curves and, conversely, mixing up a hand sample from various sands, silts and clays, in order to sketch the possible grading curves before normal sieve analysis gives the correct particle-size analysis.

Criteria of grading
There are two figures which may be quoted in order to provide an approximate description of a material. By using these regularly, the

engineer can assess subjectively the type of material in question. The figures are:

Effective size, which is the maximum particle-size of the smallest 10 per cent of the sample. This figure is obtained by reading off from the chart the size at which the grading curve cuts the horizontal line through 10 per cent.

Uniformity coefficient, which is the ratio of the maximum size of the smallest 60 per cent to the effective size. This figure is calculated by finding the size at which the grading curve cuts the horizontal 60 per cent line, and dividing by the effective size.

In using these figures, the engineer must realize their significance. Ninety per cent of any sample shows particle-sizes bigger than the effective size. As the uniformity coefficient increases above unity (the minimum possible) the samples show an increasing range of size of particle. A soil with an effective size of 0·002 mm and a uniformity coefficient of 1000 would possess 10 per cent of clay particles, but would have a very wide range of different sizes, 40 per cent being larger than 0·2 mm. It is unusual to find soils with this wide grading.

Soils D and E of Fig. 2.2 show the following characteristics:

	Largest size of the smallest (mm)		Effective size	Uniformity coefficient
	10%	60%		
D	0·11	0·20	0·11	1·8
E	0·24	1·06	0·24	4·4

Any uniformity coefficient approaching 1·0 (the minimum theoretically possible) shows a high degree of uniformity.

2.5 SECONDARY CLASSIFICATION FOR FINE GRAINED SOILS

It is possible to find two fine-grained soils with closely approximating grading curves and yet showing widely divergent engineering properties and characteristics. This divergence is caused by the varying properties of the clay minerals in the soil, and the different effects shown by the two materials when the moisture content varies. It is therefore necessary to devise a secondary classification for fine-grained soils.

Consistency limits

Fine-grained soils have been formed in nature by the gradual deposit of solid particles from suspension in water. In the process of settlement, consolidation and drying out the material passes through several well-defined stages:

    (i) suspension in liquid;
   (ii) viscous liquid;
  (iii) plastic solid;
  (iv) semi-plastic solid;
   (v) solid.

These stages are shown diagrammatically in Fig. 2.3. The changes from one stage to another are accompanied by important changes in physical properties.

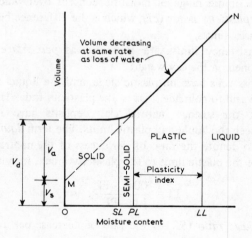

Fig. 2.3 Consistency limits.

The moisture contents at which the soil passes from one stage to the next are known as *consistency limits*. The significance of these limits was first demonstrated by Atterberg, a Swedish soil scientist, who devised simple tests for finding them. From these the present standard tests, which are described in the succeeding paragraphs, have been evolved.

The *moisture content* of a soil is defined as the ratio of the weight of water to the weight of solid in a given volume. The symbol *m* will be used to denote the moisture content expressed as a ratio, i.e.

$$m = \frac{weight\ of\ wet\ sample - weight\ of\ dry\ sample}{weight\ of\ dry\ sample}$$

but numerical values are usually quoted as percentages.

The *liquid limit (LL)* is the minimum moisture content at which the soil will flow under its own weight. As obtained from the standard test it is defined as the moisture content at which twenty-five taps in the liquid limit device will just close a groove in a sample of soil.

The *plastic limit (PL)* is defined as the minimum moisture content at which the soil can be rolled into a thread 3 mm in diameter without breaking.

The *shrinkage limit (SL)* is the moisture content at which further loss of moisture does not cause a decrease in the volume of the soil.

A measure of the range of moisture content over which a soil is plastic is the *plasticity index (PI)*, which is the difference between the liquid and plastic limits.

These consistency limits are expressed as *percentage moisture content* reckoned on the dry weight.

Cohesionless soils have no plastic stage, and the liquid and plastic limits may be said to coincide, that is, the plasticity index is zero.

Except at the surface, natural clay deposits have a moisture content between the liquid and plastic limits. The term *liquidity index (LI)* is used to denote the ratio of the excess of the natural moisture content above the plastic limit to the plasticity index. In symbols

$$LI = \frac{100m - PL}{LL - PL}.$$

The *shrinkage ratio (SR)* is the rate of decrease per unit volume with decrease in moisture content. The volume is expressed as a percentage of the dry volume and the moistue content as a percentage of the dry weight. It is therefore equal to the slope of the line *MN* (Fig. 2.3) representing the relationship between volume and moisture content.

If the straight line in Fig. 2.3 is produced back to the axis, which is the ordinate of zero moisture content, the point M represents the state of the sample if the loss of water would be continued indefinitely without the formation of air voids. The volume V_s, therefore, represents the volume of solids in the dry state and V_a the volume of air voids included in the dried sample. V_d is the total volume of the dried sample.

Some examples of the values of the consistency limits for different types of soil are given in Fig. 2.4. As the predominant particle-size decreases, both the liquid and the plastic limits increase, the former at the greater rate. It may thus be said that, in general, the plasticity index increases as identification passes from silts to the finer-grained clays.

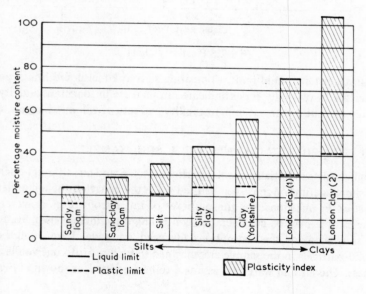

Fig. 2.4 Liquid and plastic limits for various soils.

In one of the systems of classification, therefore, a cohesive or fine-grained soil is identified by its *grading analysis* and by its principal consistency limits: *liquid limit* and *plastic limit*.

The use of the consistency limits is shown in Fig. 2.5. The group symbol in the Unified Classification System (see Table 2.3) is indicated by the point obtained by plotting the plasticity index of the

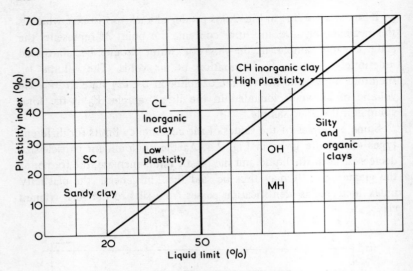

Fig. 2.5 Plasticity chart.

soil against its liquid limit. Comparing soils at equal liquid limits, an increase in plasticity index indicates an increase in toughness and dry strength, but a decrease in permeability and rate of volume change.

2.6 THE PHYSICAL STATE OF A SOIL SAMPLE

Soil consists essentially of solid particles, water, and air, although such a simple definition must be qualified when the behaviour of soil under load is in question. For the purposes of this introduction, however, the sample may be considered as containing these three phases. Study is simplified and the conceptions advanced are made clearer if all the solid particles in the sample are imagined to be fused into one voidless mass. The air and water are assumed to be separated, as shown in Fig. 2.6.

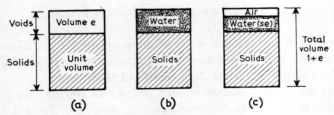

Fig. 2.6 Model soil sample.

The soil sample can be in one of three conditions:

(i) The voids can be *completely dry* and contain only air in the voids. This is the state of a sample artificially dried in an oven (Fig. 2.6a).

(ii) The voids can be *completely saturated* with water. Such a condition is common on engineering sites (Fig. 2.6b).

(iii) The sample can be *partially saturated,* the voids containing air as well as water. Such soils are known as *unsaturated soils* (Fig. 2.6c).

Mass and volume

To understand the physical state of a soil sample the student should be able to determine masses, volumes, and densities for each constituent – solid, water, and air – and for the total sample. When working in metric units, the volume of water in millilitres is defined by the same numerical figure as its mass in grams, while the mass of one millilitre of solid particles in a voidless mass is represented by the same figure as the specific gravity.

Taking the specific gravity of water as unity and that of the solid particles as G_s (usually about 2·7) the relationships given in Table 2.4 are developed from the conditions shown in Fig 2.6.

Table 2.4 *Model soil sample (Fig. 2.6)*

Sample		State					
		Dry		Saturated		Unsaturated	
		Mass (g)	Volume (ml)	Mass (g)	Volume (ml)	Mass (g)	Volume (ml)
Voids	Air	zero	e	zero	zero	zero	$(1-s)e$
	Water	zero	zero	e	e	se	se
Solids		G_s	unity	G_s	unity	G_s	unity
Total		G_s	$1+e$	G_s+e	$1+e$	G_s+se	$1+e$

Void ratio and porosity

The *void ratio, e,* is the ratio of the *total volume* of voids (whether filled with water, air, or a mixture of the two) to the *volume of solids*. In Table 2.4 the volume of solids is unity and the *void ratio* is thus e.

A less important relationship, which can be easily determined from the void ratio, is the *porosity*. This is defined as the ratio of the volume of voids to the total volume. From Table 2.4 this gives

$$\text{Porosity}, n = \frac{e}{1 + e}.$$

Both the void ratio and the porosity are dimensionless quantities.

Densities

The density of a soil sample and of soil *in situ* plays a very important part in the study of soil behaviour. It is recorded in g ml^{-1} or Mg m^{-3}, both of which are recorded by the same numerical value.

The *mass density of water* is denoted by the symbol ρ_w. Its value is unity when using the metric system.

The *mass density of soil,* or mass per unit volume, may be either:

    Bulk density, ρ, when the soil is in its natural state, whether saturated or unsaturated; or
    Dry density, ρ_d, when the sample has been dried at $105°$ to $110°\text{C}$ to drive off the water.

The density of soil is calculated from Table 2.4 by dividing the value in the *mass* column by that in the *volume* column, the result giving the required *mass* per unit volume.

$$(\text{Mg m}^{-3} \text{ or g ml}^{-1})$$

Dry density or density of *dry* sample	$\dfrac{G_s}{1 + e}\,\rho_w$
Bulk density of *saturated* sample	$\dfrac{G_s + e}{1 + e}\,\rho_w$
Bulk density of *unsaturated* sample	$\dfrac{G_s + se}{1 + e}\,\rho_w$

where $\rho_w = 1$. In the above expressions, s denoted the degree of saturation, that is the ratio of the volume of water in the sample to the total volume of voids.

The value of the density in Mg m^{-3} is numerically the same when expressed in $g\,ml^{-1}$. The latter is a more convenient unit for laboratory use.

In problems of soil mechanics we are often concerned with the *unit weight*, or weight per unit volume, sometimes described as the *weight density*. Weight densities are denoted by the symbol γ in contrast to mass densities for which the symbol ρ is used. Thus γ is the bulk weight density, γ_d the dry weight density and γ_w the weight density of water.

The force applied to a mass of 1 kg under the action of gravity (i.e., its weight) is $9 \cdot 81$ N. The *unit weight* or *weight density* in $N\,m^{-3}$ is therefore equal to the *mass density* in $kg\,m^{-3}$ multiplied by $9 \cdot 81$. The weight density of water γ_w is thus $9 \cdot 81\,kN\,m^{-3}$, and the approximation $10\,kN\,m^{-3}$ is often used.

When soil is submerged, its weight is partly supported by the buoyancy of the water displaced. The submerged weight density is therefore

$$9 \cdot 81(\rho - 1) \text{ or } 9 \cdot 81 \left(\frac{G_s - 1}{1 + e}\right).$$

Moisture content
It is important to know the moisture content or proportion of water in the soil sample. This is measured by obtaining from Table 2.4 the ratio of the *weight of water* in the sample to the *weight of solids*. The sample is weighed, dried at a temperature of $105°$ to $110°$ C, and weighed again. The loss in weight represents the weight of water in the sample. The moisture content is expressed as a dimensionless ratio, but is more frequently quoted as a percentage.

Density, void ratio and moisture content
Moisture content is easily measured, and from that figure the value of the void ratio can be determined. From Table 2.4, the mass of water in the unsaturated sample is se and is related to the moisture content by the expression:

$$\text{Moisture content } w = \frac{\text{weight of water}}{\text{weight of solids}} = \frac{se \times 9 \cdot 81}{G_s \times 9 \cdot 81}$$

Thus

$$se = wG_s.$$

If the soil is saturated, s is unity and $e = wG_s$.

Substituting wG_s for se in the expression for bulk density, Bulk mass density,

$$\rho = \frac{G_s + se}{1 + e} = \frac{G_s + wG_s}{1 + e}.$$

Dividing this by the dry mass density $\rho_d = (G_s / 1 + e)$ we get

$$\rho = \rho_d(1 + w).$$

Degree of saturation

The degree of saturation of a particular sample is obtained by determining the bulk density and the moisture content. Two equations can then be solved to eliminate the unknown void ratio and determine the degree of saturation. The values of ρ and w are known, and G_s is measured or assumed to be 2·7.

$$\rho = \frac{G_s + se}{1 + e}\,\rho_w \text{ and } se = wG_s$$

$$(\rho_w = 1 \text{g ml}^{-1} \text{ or } 1 \text{ Mg m}^{-3}).$$

Combining these

$$s = \frac{wG_s\rho}{G_s(1 + w) - \rho}.$$

2.7 APPLICATION OF THE WORK OF CHAPTER 2

Application 2A

The moisture content of a sample of unsaturated clay is 32·3 per cent. Find the void ratio and the bulk and dry densities (specific gravity of soil particles = 2·70).

The problem is conveniently worked in tabular form:

	Relative masses	Relative volumes	Volume (m³)
Water	32·3	32·3 = 32·3	0·466
Solids	100·0	100/2·7 = 37·0	0·534
	132·3	69·3	1·000

The volumes in the last column are obtained by dividing 1·000 in the proportions of 32·3 to 37·0.

The void ratio $e = \dfrac{32 \cdot 3}{37 \cdot 0} = 0 \cdot 87$.

The dry density = $0 \cdot 534 \times 2 \cdot 7 = 1 \cdot 44$.
Weight of water per m^3 = $0 \cdot 4666 = \underline{0 \cdot 47}$
Thus, *bulk density* = $1 \cdot 91$ Mg m^{-3}.

Application 2B
The bulk density of a sample of clay is 1·95 Mg m^{-3}, its moisture content is 25·3 per cent, and the specific gravity of the particles is 2·7. Find the degree of saturation.

	Relative masses	Mass density (Mg m^{-3})	Volume (m^3)	
Air	0·0	0·00	(by difference)	0·03
Water	25·3	0·39	0·39/1	0·39
Solids	100·0	1·56	1·56/2·7	0·58
	125·3	1·95		1·00

The total of the second column must be 1·95; thus 25·3 and 100·0 of the first column must be reduced in the ratio 1·95/125·3.

The value of the degree of saturation s can be determined in two ways:

$$\frac{\text{Volume of water}}{\text{Volume of voids}} = \frac{0 \cdot 39}{1 \cdot 00 - 0 \cdot 58} = 0 \cdot 93 \text{ or } 93\%$$

from

$$s = \frac{WG_s \rho}{G_s(1 + w) - \rho}$$

$$= \frac{\dfrac{25 \cdot 3}{100} \times 2 \cdot 7 \times 1 \cdot 95}{2 \cdot 7(1 \cdot 253) - 1 \cdot 95} = 0 \cdot 93 \text{ or } 93\%.$$

Application 2C
Assuming that each of the three soils described below is of a substantial depth, describe their characteristics, and put them in order of suitability as a foundation material. The consistency limits are:

Soil	Liquid limit (%)	Plastic limit (%)
F	27	12
G	5	3
H	65	42

(i) First, find the *Plasticity Index* by obtaining the difference between the *LL* and *PL*. These result in *F*, 15 per cent, *G*, 2 per cent and *H*, 23 per cent.

(ii) From the *plasticity chart*, Fig. 2.5, the soils fall into categories as follows: *F* is *CL*, *G* is below any of the cohesive categories and is thus the next higher in the *Unified Classification* and is probably *SC*, and *H* is classified as *OH*.

(iii) Before the capability of each soil as a foundation material can be decided upon, further tests and calculations must be made as is shown in later chapters, but a preliminary judgment can be made from the classification. Study the *Unified Classification* Table 2.3 to arrive at the following conclusions.

(iv) *Soil F:* This is an inorganic clay of low plasticity, perhaps containing gravel or sand, as could be decided on visual inspection. If it is dense it is probably an acceptable foundation material, and certainly worthy of further consideration.

Soil G: This is a mixture of sand and clay, tending towards the sand range rather than the clay. It will show a good many of the characteristics of a granular soil (see the table of comparison in Section 2.2). Worth further consideration.

Soil H: A highly plastic and organic soil, probably containing peat, is quite unsuitable as a foundation material, especially when it is of a *substantial depth* and thus cannot be dug out and replaced with granular fill.

The best of these three is probably *F*, although it would show some long-continued settlement. The settlement of *G* would be completed in a shorter time, but such sand/clay mixtures may not be homogeneous. Further inspection by pits and *in situ* and laboratory tests would give more data on which to make a decision.

CHAPTER 3

Pore-water in soils

This chapter deals with water in soils when the water is at rest under equilibrium conditions. The following chapter considers the conditions existing when the water is affected by an unbalanced system of forces, and flow through the soil is taking place.

3.1 PORE-PRESSURE AND THE WATER TABLE

Water in soil affects its engineering properties and plays a very important role in all soil mechanics problems. The effects caused by the presence of water, whether at rest or moving through the pores of the soil, must be thoroughly understood.

The water table
Water in soil obeys the laws of hydraulics and, when at rest, exerts a pressure at any point consistent with the depth of water above that point. This is as true for water within the pores of the soil as for free-standing water. Atmospheric pressure is accepted as the datum for pressure-head measurements, and the level at which the pressure in the water is atmospheric is called the *water table* or, more correctly, the *phreatic surface*. This area, over which the pore-pressure is zero, is normally found to follow the ground surface, but at a slightly flatter slope. There are, however, discontinuities in the water table in certain geological conditions, and there may be more than one level on a single vertical section where the pore-pressure is zero and a water table exists. Small areas of zero pore-pressure lying above the general water table for the district are known as *perched water tables*. In a well or borehole the water table is defined by the visible surface level of the

35

water. This visible surface is formed at once when the well or borehole is driven below the water table in pervious or water-bearing soil. In cohesive soils, however, a visible surface will not be formed at once, for time is necessary for the pore-pressure in the soil to force water out into the borehole and attain equilibrium conditions of pressure.

Positive and negative pressure
Below the water table the pore-water is under normal hydrostatic head, the pressure increasing with depth as in free-standing water. Above the water table the soil may still contain considerable quantities of water. This fact explains why it is incorrect to think of the water table as a physical upper surface to the pore water. It is more correct to think of it as a surface of zero pressure between the positive pore-pressure below and the negative pore-pressure, or *soil suction,* above.

Above the water table the water is maintained in place, against the force of gravity, by capillary attraction and surface tension in the voids. Capillary water may exist in three forms:

(i) *capillary saturation* where the voids are completely filled with water;

(ii) *partial saturation.* where there is some air, but the water in the voids is continuous;

(iii) *contact moisture,* where there is more air, and the water is discontinuous, existing in separate aggregations.

The attraction exerted between the particles by the capillary water is known as *soil suction,* and has a pressure which is less than atmospheric. When the capillary water is in a state of equilibrium, the suction exerted by the water between two particles is balanced by the pressure between the particles. Capillary water has, of course, a marked effect on the cohesion of the soil. The greater the soil suction, the greater the cohesion. Soil suction increases with decrease in the size of voids and capillary channels; it follows that fine-grained soils have greater cohesion than coarse-grained.

A portion of the capillary water exists in the form of very thin films surrounding the individual grains of soil. Water in this form, known as *adsorbed* water, has properties differing from those of ordinary water. It has greater viscosity, greater surface tension and a higher boiling point. Adsorbed water cannot be removed by air drying at ordinary temperatures. The amount of adsorbed water present in soil is usually assumed to be the difference between the weight of a sample when air-dried, and that when oven-dried at 105 to 110° C.

3.2 EFFECTIVE PRESSURE AND PORE-PRESSURE

The total pressure on a horizontal area within a soil mass is made up of two types of pressure: that transmitted through the chains of soil particles, and that transmitted through the water – the pore-pressure. Fig. 3.1 gives a diagrammatic representation of how the weight of the soil, plus any superimposed load, is transmitted. The load transmitted by intergranular pressure is applied at E, E and the pore-pressure applied also to the base LL acts quite independently through the voids. (Fig. 3.1).

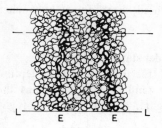

Fig. 3.1 Action of effective and pore-pressures

When the soil is stressed in an engineering structure, only the intergranular contacts induce any resistance to deformation and failure. The water, having zero shear resistance, is ineffective or neutral and the pore-pressure, u, may thus be called the *neutral pressure*. The intergranular pressure is known as the *effective pressure*, since it is 'effective' in mobilizing shearing resistance, and in causing volume change and deformation. The effective pressure, σ', is generally defined as the difference between the total stress σ and the pressure, u, set up in the pore fluid, as expressed by the equation

$$\sigma' = \sigma - u.$$

Effective stress may thus be described as the average intergranular force per unit area of plane and must not be confused with the individual contact pressures between the grains.

If a the area of contact between the particles per unit area of plane, is taken into account, the total stress is

$$\sigma = \sigma' + (1 - a)u$$

that is, the effective stress

$$\sigma' = (\sigma - u) + au.$$

It can be shown that for compressible soils the volume change depends only on $\sigma - u$ and is not affected by a. For rocks with rigid structure, however, the usual expression $\sigma' = \sigma - u$ is not strictly correct. As regards shear strength, a is generally very small for saturated and nearly saturated soils and may be neglected.

Hence the general assumption may be made that for all soils except where the saturation is less than about 80 per cent or where the structure is solid, the simplified expression for effective stress, $\sigma' = \sigma - u$, may be used, both for volume change and for shear strength.

Effective pressure under static equilibrium
Referring to Fig. 3.2, the total pressure at depth d equals the weight of a column of dry sand of height d_1 plus that of a column of saturated sand of height $d - d_1$. This pressure is given by

$$\sigma = \gamma_d d_1 + \gamma(d - d_1).$$

The effective pressure at depth d is therefore given by

$$\sigma' = \sigma - u = \gamma_d d_1 + \gamma(d - d_1) - \gamma_w(d - d_1)$$

$$= \gamma_d d_1 + (\gamma - \gamma_w)(d - d_1).$$

These conditions apply to coarse sands when the water table is also the upper limit of water in the soil. Where, however, the particles are very fine, the soil above the water table may be saturated for all or part of its depth by capillary water held in suspension by soil suction. The capillary water is held suspended above the water table and thus does not alter the values of the pore pressure below the water table. The effective pressure is, however, increased by the weight of water, since this is held in suspension by suction and can be visualized as 'hanging' on the soil particles above the water table. The effective pressures below the water table are thus increased by the difference $d_1(\gamma - \gamma_d)$. Above the water table the effective pressures are

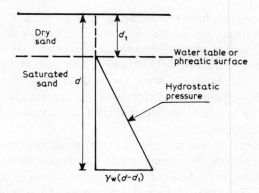

Fig. 3.2 Pore-pressure from static head

increased by $z(\gamma - \gamma_d)$ where z is the depth below the surface to the level at which the pressures are measured, assuming that the whole of the soil above the water table is saturated with capillary water.

Pore-pressure coefficients

In problems relating to the stability of earth dams and other large masses of soil it has been found useful to express the relation between the change in pore-pressure and the changes in the principal stresses by the use of two constants, A and B, known as *pore-pressure coefficients*. These coefficients can be measured experimentally in the triaxial compression test.

Coefficient B: Consider first a sample of soil subjected to an all-round pressure σ. If this pressure is increased by an amount $\Delta\sigma$ and drainage is prevented, there will, in general, be an increase of pore-pressure Δu. The ratio $\Delta u/\Delta\sigma$, denoted by B, depends on the compressibility of the soil structure as a whole, relative to that of the fluid contained in the voids. In saturated soil the voids are filled with water which is relative incompressible, and therefore $B = 1$. In perfectly dry soil the air is highly compressible and B is zero. For partially saturated soils B has some value between 0 and 1.

Coefficient A: Now consider an increment of pressure $\Delta\sigma$ in one direction only. If the soil structure is regarded as an elastic solid with linear stress/strain characteristics, the volumetric strain produced by a pressure increment, in one direction only, is one-third of that due to an all-round increment of pressure of the same magnitude. Hence the theoretical increase of pore-pressure $\Delta u = \frac{1}{3} B \Delta \sigma$. As the stress/strain characteristics of the soil skeleton are not linear, the factor $\frac{1}{3}$ must be replaced by a coefficient A which can be determined experimentally. This coefficient varies considerably with the type of soil from +1 or even more for normally consolidated clays of high plasticity, to $-0 \cdot 5$ for overconsolidated clays.

An element of soil is originally in equilibrium under a two-dimensional horizontal pressure σ_3 and a vertical pressure σ_1. Due to external loading these principal stresses are increased by $\Delta \sigma_3$ and $\Delta \sigma_1$ respectively. The stress changes may be considered as taking place in two stages, namely, an equal all-round increment $\Delta \sigma_3$ and a deviator stress $(\Delta \sigma_1 - \Delta \sigma_3)$ in the direction of σ_1. The resultant change in pore-pressure is thus the combination of these two stages and may be expressed as

$$\Delta u = B[\Delta \sigma_3 + A(\Delta \sigma_1 - \Delta \sigma_3)]$$

Coefficient $\bar{B}$: For practical purposes the two coefficients can be combined to give the simpler expression

$$\frac{\Delta u}{\Delta \sigma_1} = \bar{B}.$$

This is evaluated as follows:

$$\frac{\Delta u}{\Delta \sigma_1} = B \left(\frac{\Delta \sigma_3}{\Delta \sigma_1} + A - A \frac{\Delta \sigma_3}{\Delta \sigma_1} \right)$$

$$= B \left[(1 - A) \frac{\Delta \sigma_3}{\Delta \sigma_1} - (1 - A) + 1 \right]$$

$$= B \left[1 - (1 - A) \left(1 - \frac{\Delta \sigma_1}{\Delta \sigma_3} \right) \right] = \bar{B}.$$

The expression for $\bar{B}$ involves the ratio $\Delta \sigma_3 / \Delta \sigma_1$, which depends on the degree of lateral confinement. For rigid confinement, as in the consolidation test, $\sigma_3 = \sigma_1$ and $\bar{B} = 1$. On the assumption that σ_3 and σ_1 change proportionally, $\Delta \sigma_3 / \Delta \sigma_1$ is equal to σ_3 / σ_1. If the soil is easily displaced laterally, the ratio σ_3 / σ_1 is equal to the coefficient of earth pressure at rest. The usual condition is between these two extremes and σ_3 / σ_1 varies from about $0 \cdot 4$ to $0 \cdot 75$.

3.3 SOIL SUCTION

Soil suction may be defined as the negative pressure by which water is retained in the pores of a sample of soil when the sample is free from external stress. A pressure deficiency (below atmospheric pressure) may also exist in soil subjected to certain stress regimes associated with particular loading conditions. Such pressure deficiency in a stressed soil is called negative pore-pressure to distinguish it from the natural suction occurring in unstressed soil.

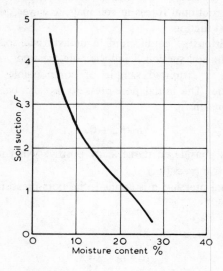

Fig. 3.3 Suction/moisture content relation

As moisture content decreases, soil suction increases and may become very large with oven-dry soil. Fig. 3.3 shows a typical relationship between moisture content and soil suction. Similar curves have been produced showing the increase in shear strength which accompanies an increase in soil suction. The suction/moisture content relationship is not unique. If the soil, initially dry, is gradually wetted, the curve is slightly different from that obtained under drying conditions.

Soil suction is generally measured as centimetres of hydraulic head, and as this value may become very large as the soil dries out, it is usual to employ the common logarithm of the head as a measure of soil

41

suction. The special scale devised for the measurement of soil suction was developed by Schofield and is known as the *pF* scale. It is analogous, in presentation of conditions, to the Sorensen acidity scale of *pH*. A *pF* value of 2, therefore, corresponds to a negative hydraulic head of 10^2 or 100 cm, that is to 1 m, and the suction pressure is 9.81 kN m^{-2}.

There is some uncertainty about the true nature of soil suction. It can be argued that it represents no more than a capillary phenomenon and is therefore, merely a function of the soil matrix. Osmotic, adsorptive and frictional forces in soil may, however, play their parts in producing soil suction.

Under conditions of equilibrium in undisturbed soil the relationship between the soil suction S and the pore-pressure u may be found by considering a saturated sample of compressible soil with no external pressure. The initial pore-pressure u_0 just balances the soil suction and therefore

$$u_0 + S = 0.$$

(Soil suction is considered numerically positive when it is actually a suction or negative pressure.)

If an external pressure σ is applied, the pore-pressure increases to

$$u = \sigma + u_0 = \sigma - S$$

and the effective stress

$$\sigma' = \sigma - u = S.$$

A more general relation between S and σ is

$$S = \alpha\sigma - u$$

where α is the fraction of the total normal pressure which is effective in changing the suction.

For saturated clays, which are compressible, $\alpha = 1$, the applied pressure being taken by the water. For incompressible soils such as compact sand, or for permeable rock with a rigid structure such as chalk, α is zero, since any applied pressure is resisted by the soil structure and not by the pore-water. For soils of low compressibility α has a value intermediate between 0 and 1. For example, α may be about 0.5 for a silty clay or 0.15 for a sandy clay.

The general equation, $S = \alpha\sigma - u$, is applicable only when the soil is saturated or nearly so.

3.4 EQUILIBRIUM MOISTURE CONTENT

If prescribed conditions of pressure and drainage are applied to a soil (especially a cohesive soil) the mass will attain, in time, a moisture content profile which remains constant under the specified external conditions, and is thus called a condition of *equilibrium moisture content*. So long as the soil remains saturated, any change in the applied conditions, resulting in the addition or removal of water, causes a change of volume of the soil equal to the volume of moisture added or subtracted.

It is important to be able to evaluate the equilibrium moisture content at levels below the surface for different states of surface loading and for different levels of the water table. A road slab, for example, laid on a clay subgrade in dry weather may subsequently be raised by the expansion of the clay as it reaches the equilibrium moisture content. In a saturated soil the moisture content reached under equilibrium conditions is proportional to the void ratio, and this depends on the effective pressure. The effective pressure, in turn, is related to the positive or negative pressure in the pore-water which fills the voids.

Determination of equilibrium moisture distribution

The estimation of the equilibrium moisture content can be started from either end of the chain of inter-related quantities, that is, either from the relationship between soil suction and moisture content, or from that between effective pressure and void ratio. Whichever method is adopted, the first step is to find the effective pressure at various depths.

The intergranular or effective pressure, σ', is the algebraic difference between the total pressure applied by the superincumbent load and the pressure in the pore-water. The latter is controlled by the position of the point in question in relation to the phreatic surface. Under equilibrium conditions the effective pressure at any depth is the same as the suction pressure corresponding to the moisture content at that point.

Effects of variation in height of water table and of superimposed load. If the water table or phreatic surface is at a depth d below the surface the effective pressure at depth z is

$$\sigma' = \gamma z - \gamma_w(z - d) = (\gamma - \gamma_w)z + \gamma_w d.$$

If z is greater than d, the effective pressure σ' is less than the pressure due to the bulk density because of the uplift exerted by the gravitational water. If z is less than d, the effective pressure is greater than the pressure due to the bulk density alone by an amount equal to the negative pore-pressure. The effect, therefore, of lowering the water table from the surface to a depth d below the surface is to increase the soil suction and the effective pressure at all depths by a constant amount $\gamma_w d$.

A superimposed load exerts its effect in one of two ways. If the area of contact between the load and the surface is small, then the compressive stress decreases with increase of depth. If, however, the surface covered by the load is large in comparison with the depths under consideration, as is usual for the dead loads of road and runway pavements, the pressure exerted by the superimposed load can be considered constant and uniform at any given depth.

Equilibrium moisture distribution from soil suction
Since the moisture content of the soil varies with the depth, the bulk density also varies. As a first approximation a constant value is assumed for the density. Using this value the effective pressures at a series of depths are calculated. Under conditions of equilibrium the soil suction is equal to the effective pressure and if a suction/moisture-content relation for the soil such as that shown in Fig. 3.3 is available, values of the moisture content at the various depths can be calculated. These will not be quite correct, since the variation in density with depth was not taken into-account. Using the values of m obtained from the first approximation, corrected values for the density can be worked out, and using these the procedure can be repeated to give a more accurate estimate of the equilibrium moisture distribution.

Equilibrium moisture distribution from void ratio
The second method of determining the equilibrium moisture distribution depends on the pressure/void-ratio relationship for the soil, as obtained from a consolidation test. Here again a constant bulk density is first assumed, from which the pressures at various depths are calculated. The corresponding void ratios are read off from the p/e curve and the moisture content calculated for the saturated soil, from $e = mG_s$. These values enable corrected densities to be worked out

and by repeating the process a closer estimation of the equilibrium moisture distribution can be obtained.

In most practical problems, whichever method is employed, the calculation of the second approximation is unlikely to be necessary, provided that a reasonable value has been assumed for the average density.

3.5 FROST HEAVE AND PERMAFROST

When water freezes in a void in a saturated soil mass, so that the resulting ice fills the void, further freezing can take place in two ways:

(i) the growth of ice can continue in neighbouring voids by the formation of minute crystals growing from the original one; or

(ii) the original crystal can continue to grow in size without extension of the freezing to the surrounding pores.

In certain soils the second of these conditions may prevail, with consequent serious heaving of the surface soil due to the continued growth of crystals into ice lenses. The conditions required are: some large voids in which ice lenses may begin to form as well as very small pores (around 0·002 mm) in which freezing will not take place [as in (i) above] until the pressure on the original ice lens reaches a limiting value called the *heaving pressure*. This limiting value of pressure cannot well be mobilized by a normal road surface, hence much frost damage is seen on roads. The smaller the pores in the soil, the greater is the pressure required to stop the local and harmful formation of ice lenses and propagate relatively harmless freezing throughout the pores of the soil mass. Silts exhibit a structure in which frost heave is likely to take place; variation in pore size from larger voids in which individual crystals may start to grow to voids of such small dimensions that the associated heaving pressure is very high.

Apart from removing the silt and replacing it with sand having uniformly large voids (as is done in Sweden), there are two precautions which can be taken. The water table may be depressed by good drainage, or the surface may be loaded heavily in order to allow of resistance to the mobilization of a large heaving pressure.

A phenomenon which occurs in the Arctic regions, and also in the soil underlying cold stores, is *permafrost,* a permanently frozen soil. If the soil is of the right structure, and sufficient surface load is not available, frost heave may be the result.

3.6 APPLICATION OF THE WORK OF CHAPTER 3

Application 3A

A stratum of sand, 2·4 m thick, overlies a stratum of saturated clay as shown in Fig. 3.4. The ground water level (water table) is 0·9 m below the surface. Assuming no capillary water above the water table, calculate total and effective pressures at A, B and C and draw the pressure diagram. The specific gravity of particles and the void ratio for the sand are 2·65 and 0·5, respectively, and for the clay are 2·70 and 0·9. Mass density of water is $\rho_w = 1$ Mg m^{-3} and weight density is $\gamma_w = 9·81$ kN m^{-3}.

(i) Mass density

$$\text{of dry sand} = \frac{G_s}{1+e} \rho_w$$

$$= 2·65/1·5 = 1·77 \text{ Mg m}^{-3}$$

$$\text{of saturated sand} = \frac{G_s + e}{1+e} \rho_w$$

$$= (2·65 + 0·5)/1·5 = 2·10 \text{ Mg m}^{-3}$$

$$\text{of saturated clay} = \frac{G_s + e}{1+e} \rho_w$$

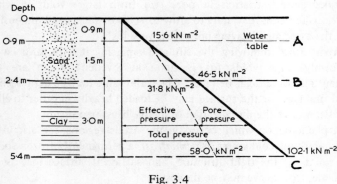

Fig. 3.4

$$= (2 \cdot 70 + 0 \cdot 9)/1 \cdot 9 = 1 \cdot 89 \text{ Mg m}^{-3}.$$

(ii) Total pressure

at A = 0·9 m dry sand

$$= 0 \cdot 9(1 \cdot 77 \times 9 \cdot 81) = 15 \cdot 6 \text{ kN m}^{-2}$$

at B = 15·6 + 1·5 m saturated sand

$$= (15 \cdot 6 + 1 \cdot 5 \times 2 \cdot 1 \times 9 \cdot 81) = 46 \cdot 5 \text{ kN m}^{-2}$$

at C = 46·5 + 3 m saturated clay

$$= (46 \cdot 5 + 3 \times 1 \cdot 89 \times 9 \cdot 81) = 102 \cdot 1 \text{ kN m}^{-2}.$$

(iii) Pore pressure

at A = 0
at B $= 1 \cdot 5 \gamma_w = 1 \cdot 5 \times 9 \cdot 81 = 14 \cdot 7 \text{ kN m}^{-2}$
at C = 4·5 $= 4 \cdot 5 \times 9 \cdot 81 = 44 \cdot 1 \text{ kN m}^{-2}$

(iv) Effective pressure (Total–pore-pressure)

at A =   15·6 –   0   = 15·6 kN m^{-2}
at B =   46·5 – 14·7 = 31·8 kN m^{-2}
at C = 102·1 – 44·1 = 58·0 kN m^{-2}.

Application 3B
If the 0·9 m of sand above the water table is not dry but saturated with capillary moisture, what differences in the figures obtained in Application 3A?

(i) The capillary water is held suspended in the voids of the sand above the water table, and thus increases the mass density to 2·10 Mg m^{-3}. This weight of water, however, does not add to the positive pore-pressure below the water table.

(ii) The additional weight of the suspended water increases the effective pressure at 0·9 m depth and at all points below by

$$0 \cdot 9 (2 \cdot 10 - 1 \cdot 77) 9 \cdot 81 = 2 \cdot 9 \text{ kN m}^{-2}.$$

Application 3C
If the water table in Application 3A rises to the surface, what is the effect on the pressures?

  (i) The upper 0·9 m is now saturated, but with water giving a positive pressure. The pore-pressures at all depths are thus increased by $0·9 \times 9·81 = 8·8$ kN m^{-2}.

 (ii) *Total pressures* are increased by 2·9 kN m^{-2} as in Application 3B.
At A, 18·5; at B, 49·4; at C, 105·0 kN m^{-2}.

(iii) *Pore pressures* are increased by 8·8 kN m^{-2}.
At A, 8·8; at B, 23·5; at C, 52·9 kN m^{-2}.

(iv) *Effective pressures*
At A, $18·5-8·8 = 9·7$ kN m^{-2}
At B, $49·4-23·5 = 25·9$ kN m^{-2}
At C, $105·0-52·9 = 52·1$ kN m^{-2}.

which shows the serious effect the presence of water has on effective stress

CHAPTER 4

Flow of water in soils

In Chapter 3 the pore-pressures discussed apply to water at rest. In these conditions there is no unbalanced *head* of water which would cause flow to take place. When an unbalanced head occurs, water is driven through the mass of soil and the pore-pressure at any point depends on the paths the particles follow in their passage to a region of lower pressure. The energy of water is measured by three types of *head:* position head, pressure head, and velocity head, the sum of the three representing the *total head,* a measure of its power to produce energy. It is assumed that the equation of Bernouilli is understood:

$$H = h + \frac{u}{\gamma_w} + \frac{v^2}{2g} .$$

In soils, the third of these, the velocity head, is negligible since the resistance to flow offered by the particles of soil is normally so great that the velocity is small. The position head and the pressure head u/γ_w are, however, very important and represent, together, the *total head.*

Flow of water through soil can be induced in several ways, of which the most important are:

(i) the existence of a water table or free surface on each side of a soil mass, one of these surfaces being higher than the other. This condition induces *seepage.*

(ii) The application of pressure (e.g. the weight of a building). This increases the pore pressure above its static value and induces flow.

4.1 SEEPAGE

In developing the laws governing the flow of water through soils, Darcy's law is assumed to be applicable, namely

$$q = Aki \quad \text{or} \quad v = ki$$

where q = flow in unit time through a cross-sectional area A of soil
  k = coefficient of permeability
  i = hydraulic gradient
  v = velocity of flow.

Coefficient of permeability
This quantity is defined as the rate of flow per unit area of soil under unit hydraulic gradient.

The coefficient of permeability has the dimensions of a velocity, and is usually expressed in m s^{-1} or mm s^{-1}. It represents the velocity which would produce the same rate of discharge if the water flowed through the whole area instead of through the voids.

The coefficient, usually denoted by k, is a function of:

  (i) the porosity of the soil and the shape and size of the voids;
  (ii) the density and viscosity of the fluid.

The coefficient of permeability thus applies only to a particular fluid – in our case, water – and is subject, like viscosity, to slight variation with temperature.

For granular materials, the permeability generally varies inversely with the *specific surface of the particles,* i.e. the surface area per unit weight. With saturated clays the permeability varies considerably with the moisture content.

If the void ratio of the soil is e, the ratio of the cross-sectional area

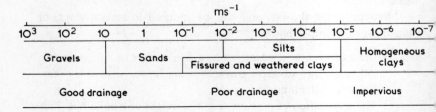

Fig. 4.1 Permeability ranges.

of the pores to that of the whole soil is $e/(1 + e)$. The true average velocity v_s with which the water travels through the soil, known as the *seepage velocity* or velocity of percolation, is therefore given by

$$v_s = \left(\frac{1 + e}{e}\right) ki.$$

Fig. 4.1 shows the average values of the coefficient of permeability, k for various soils.

Variation of permeability with depth and direction of flow

Frequently the soil mass through which seepage takes place consists of several strata with different coefficients of permeability. Even when the soil is of fairly uniform composition the permeability tends to decrease with the depth because of the increasing density caused by the weight of superincumbent strata. In many natural soil deposits the permeability in the horizontal direction is several times that in the vertical direction. An earth mass which does not conduct moisture equally well in all directions is known as *anisotropic*.

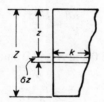

Fig. 4.2 Variable permeability.

When the permeability varies with depth, a convenient approximation is to find an equivalent coefficient of permeability k_x for horizontal flow and another equivalent value k_z for vertical flow. These coefficients, though differing from one another, are assumed to be constant throughout the depth of the soil mass under consideration. The equivalent coefficients are found as follows:

Let k be the coefficient of permeability in the horizontal direction at any depth z (Fig. 4.2). The flow in a horizontal direction through an element of unit width and thickness δz is given by

$$\delta q = ki\, \delta z$$

where i is the hydraulic gradient. Then the total flow through a

51

section of unit width and depth Z is

$$q = k_x iZ = i \int_0^Z k \, dz.$$

Therefore $k_x = \dfrac{1}{Z} \int_0^Z k \, dz$ = mean ordinate of the curve showing k against z.

Considering vertical flow, let H be the head lost over a depth Z, let δh be the head lost in passing through an element of thickness δz and let the permeability of this element in the vertical direction be k. Then

$$v = k \frac{\delta h}{\delta z}.$$

To find the equivalent permeability k_z put

$$v = k_z \frac{H}{Z} = k \frac{\delta h}{\delta z}.$$

Therefore

$$\frac{dh}{k_z} = \frac{H \, dz}{Zk}.$$

Integrating between limits 0 and H and 0 and Z,

$$\frac{H}{k_z} = \frac{H}{Z} \int_0^Z \frac{dz}{k}.$$

Hence $1/k_z = (1/Z) \times$ area of curve of $1/k$ plotted against z, or k_z = reciprocal of mean value of $1/k$. In anisotropic soil this curve must be plotted from the values of the vertical permeability at various depths.

Flow equations
Consider an elementary block of soil (Fig. 4.3) of width δx, height δz, and of unit thickness (perpendicular to the plane of δx and δz). Let the velocity of flow be in the plane of the x- and z-axes, and let the components of velocity at entry to the element be v_x and v_z. The velocities at exit from the element will therefore be respectively,

$$v_x + \frac{\partial v_x}{\partial x} \delta x \quad \text{and} \quad v_z + \frac{\partial v_z}{\partial z} \delta z$$

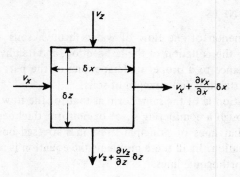

Fig. 4.3 Seepage through element of soil.

Since the quantity entering the element is equal to the quantity leaving,

$$v_z \delta z + v_z \delta x = v_x \delta z + \frac{\partial v_x}{\partial x} \delta x \delta z + v_z \delta x + \frac{\partial v_z}{\partial z} \delta z \delta x.$$

Therefore

$$\frac{\partial v_x}{\partial x} + \frac{\partial v_z}{\delta z} = 0$$

$$v_x = -k \frac{\partial h}{\partial x} \quad \text{and} \quad v_z = -k \frac{\partial h}{\partial z}.$$

Putting $kh = \phi$, a quantity which may be termed the potential, we get

$$v_x = -\frac{\partial \phi}{\partial x} \quad \text{and} \quad v_z = -\frac{\partial \phi}{\partial x}.$$

Therefore

$$\frac{\partial^2 \phi}{\partial x^2} + \frac{\partial^2 \phi}{\partial z^2} = 0.$$

This is a two-dimensional case of the well-known Laplace equation.

4.2 FLOW NETS

The phenomenon of the flow of water through soils – seepage – is governed by the equation of the last section, and its physical meaning is of importance to a proper understanding of the part flowing water plays in the stability and strength of soils.

The equation is of the same form as that for the flow of an electric current through a conducting sheet of uniform thickness. It can also be shown that lines of principal stress in a stressed body satisfy the Laplace equation. In all these problems the equation is represented by two sets of orthogonal lines.

Flow and equipotential lines

The two sets of orthogonal lines which represent the equation of flow are illustrated in Fig. 4.4. Such a diagram is known as a *flow net*. Those lines bearing arrows are the *flow lines,* which represent the paths of flow of particles of water moving through the soil. The spaces between flow lines may be called *flow channels*. The lines orthogonal to these are known as *equipotential lines*.

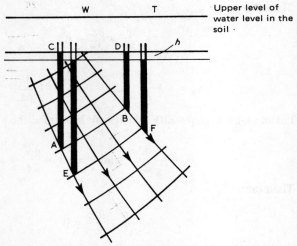

Fig. 4.4 Flow of equipotential lines.

If *piezometer tubes* (instruments for measuring pore-pressure) are inserted in the soil at *A* and *B*, two points on one of the equipotential lines, the pore-pressures, represented by the height the water would rise in the piezometer tubes, are shown to be *AC* and *BD*, defined in linear measurement. Although these pressures are different, the pore-water at A and at B has the same potential. At A there is a higher pressure but a lower position head. At B, the height of the point above datum is greater and the pore-water thus requires a lower pressure to bring the total head to the same value as at A.

Seepage pressure

At level WT all particles of water have a maximum potential. By the time it has travelled through the soil to the equipotential line AB, the water has lost potential, as measured by the difference in level between WT and CD. This lost head has been transferred (as a pressure) to the particles in the mass of soil through which flow has taken place. The pressure at any point within the soil due to this 'drag' is called the *seepage pressure*.

If pore-pressure is again measured at the next equipotential line EF along the flow path, the potential is found to be less than at AB. The difference h represents the head lost. Head is converted to a pressure by multiplying by the weight density of water. The seepage or drag applied to the particles by virtue of the movement of the water and the loss of head h is thus given by

$$\text{Seepage pressure} = h\gamma_w.$$

At this point in the flow net, let us assume that the 'square' of the net has a side of length L. If a thickness L is also imagined at right angles to the paper, the force exerted in the direction of flow is $h\gamma_w L^2$. This is applied to a block approximately L^3 in volume. Thus the seepage force applied per unit of volume is

$$\frac{h\gamma_w L^2}{L^3} = \frac{h}{L}\,\gamma_w = i\gamma_w.$$

To summarize:

(i) Seepage force per unit volume of soil is represented by *Hydraulic gradient × weight density of water* (kN/m^3)
(ii) Seepage pressure for a length of flow path is: *Head loss over length of path considered × weight density of water* (kN/m^2).

Seepage pressure has an important effect on the stability of the soil mass through which the seepage takes place. Examples of this are the conditions for quicksand, piping, and internal erosion, which are discussed later in this chapter.

Sketching of flow nets

Although the flow nets can be drawn accurately after an analysis by the relaxation method or by use of the electrical analogue or other experimental method, they are most quickly and easily produced by sketching.

The first step is to form the outline of the flow net by considering the boundary conditions. Any impermeable surface, such as the waterface and the base of a dam, or the surface of an impervious stratum of soil or rock, forms a flow line. The surfaces, at entry and exit, of the soil through which the seepage takes place, are equipotential lines.

There are thus four lines defining the boundary conditions:

   (i)   the entrance surface;
  (ii) the exit surface
       (these two are equipotential lines);
 (iii) the upper flow line;
 (iv) the lower flow line.

Referring to Fig. 4.5, which shows the flow net beneath a dam, **AB** and **CD** are respectively the entrance and exit surfaces. The upper

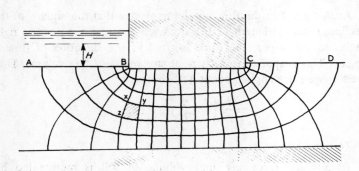

Fig. 4.5 Flow net under a dam.

flow line is the base of the dam and the lower flow line is the surface of the impervious layer.

Within the outline, a trial line is drawn from the entry surface to the exit surface. A flow net is then built up from this preliminary line. The principles to be observed in drawing the net are that the flow and equipotential lines intersect at right angles and form equal-sided figures whose continued dubdivision would result in approximate 'squares'. The first trial net will almost certainly have to be adjusted to conform to these rules and to satisfy the boundary conditions. Only by considerable personal practice can the student learn to sketch flow nets quickly and accurately.

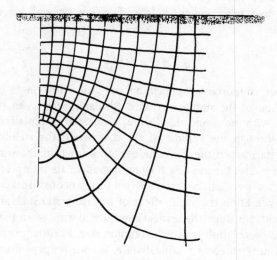

Fig. 4.6 Flow to a drain.

Another example of a flow net is illustrated in Fig. 4.6. This applies to the conditions of flow to a porous drain-pipe under constant head with a level water surface. Since successive equipotential lines represent equal drops in head, the closer the lines are together (in this example near to the drain) the greater the hydraulic gradient causing flow.

4.3 APPLICATION OF FLOW NETS

If the permeability of the soil is known we can estimate the approximate loss of water by seepage below sheet piling, under an impermeable dam, or through an earth embankment. We can also calculate the seepage pressures which will be called into play and which affect the stability of a dam.

Seepage under and through dams

If the flow net is fairly accurately plotted it can be assumed that when traversing any flow line the pressure drop between consecutive equipotential lines is the same. Assume that the 'squares' represent one unit. Suppose that the path of the water seeping from one surface to the other consists of N units and the width of the zone of flow is M units (in Fig. 4.5 $M = 6$ and $N = 15$). If H is the total difference of head, the hydraulic gradient is H/N. The flow per unit length of dam is therefore

$$q = Aki = kH \frac{M}{N}.$$

Another important application of the flow net is in the investigation of the stability of an earth dam. Even when the dam is provided with an impervious core, and zero percolation can be assumed through the volume of the dam itself, the stability of the upstream face when there is a sudden drawdown of the water in the reservoir requires the use of a flow net for adequate interpretation.

There are two critical cases for a homogeneous dam, eliminating from consideration the small effect of surface water such as rain. For the downstream slope the critical condition occurs when the reservoir is full and percolation is at its maximum rate. For the upstream slope this does not represent a critical state, for the seepage pressure then acts inwards from this slope and tends to increase the stability on the upstream side. For the upstream slope the critical condition is, rather, when the reservoir has been emptied rapidly, and when seepage is from the interior of the dam outwards, so tending to decrease the stability.

For the homogeneous dam shown in Fig. 4.7 the permeability is assumed to be uniform. The flow net for the full condition (Fig. 4.7a) must not intersect the downstream face if full safety is to be preserved; any percolation should take place completely within the dam itself. To achieve this end, the engineer may place a filter at or under the toe of the dam to allow of an easy downward passage of the water, and so deflect the flow lines in the desired direction.

Fig. 4.7b shows how, when the dam is quickly emptied, the water in the saturated material drains out either to the filter toe on the downstream side, or downwards within the upstream slope. This downward drainage within the slope calls into play a seepage pressure which assists the weight of the soil to cause a slip-circle failure. Or, by another interpretation, the neutral or pore-pressure on a likely slip surface reduces the effective or intergranular pressure, and thus decreases the shearing resistance of the slip surface.

If *SC* is a trial slip surface, the variation in pore-pressure along its length is obtained by measuring, at each of its intersections with an equipotential line, the vertical height from that intersection to the level at which the equipotential line cuts the surface of the slope (Fig. 4.7b).

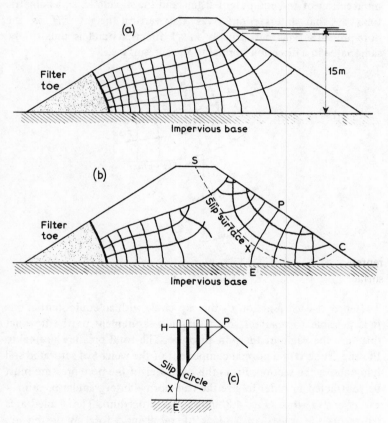

Fig. 4.7 Flow nets through dam.

The reason for this is shown in the enlarged sketch of the equipotential line *EP* (Fig. 4.7c). Taking the impervious base at datum, the head at *P* is represented by its height above the datum (position head: no pressure head). At E the head is represented by the pore-pressure corresponding to the vertical height between E and P (pressure head: no position head). An intermediate points on EP potential is made up partly of position head and partly of pressure head, but the total head available is always equal to the vertical distance between E and P.

The height of the water column in a piezometer tube represents the pore-pressure, which gradually decreases as P is approached. If X is the intersection of the equipotential line and the slip circle, a piezometric tube at that intersection gives the vertical height XH as the pore-pressure. This acts equally in all directions and is thus of the same value in a direction normal to the slip surface.

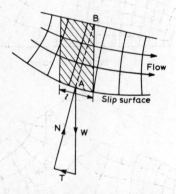

Fig. 4.8 Effective pressure on a slip surface.

Hence at each junction of the slip circle with an equipotential line it is possible to read off, by a simple measurement on the flow-net diagram, the value of the pore-pressure. The total pressure normal to the slip circle is the normal component of the weight of saturated soil lying above the section. From this total weight the pore-pressure must be subtracted in order to give the true normal intergranular pressure – the *effective stress*. Fig. 4.8 shows this procedure. The shaded area represents the weight of a block of soil giving a force W on the slip surface of length l. This gives total stress. A dotted equipotential line AB (drawn from the centre of the block on the slip surface, and parallel to the nearest line) gives for its height the pore-pressure at A. The subtraction of this value from the total normal reaction gives the effective normal stress.

Flow nets for anisotropic soil

For this condition an adaptation of the graphical method for constructing flow nets has been devised by the Swedish engineer Samsioe. It is assumed that the permeability has different values k_x and k_z in the horizontal and vertical directions respectively, but that k_x and k_z are themselves constant with depth. When the permeability varies with depth the values used should be the equivalent horizontal and vertical coefficients, k_x and k_z, determined as shown in a preceding section.

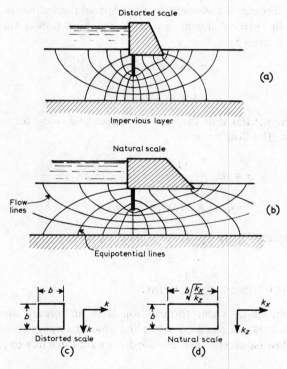

Fig. 4.9 Flow net for anisotrophic soil.

The drawing is made to a distorted scale (Fig. 4.9a), the horizontal dimensions being reduced by the factor $\sqrt{(k_x/k_z)}$ in relation to the vertical dimensions. For example, if $k_x = 4k_z$ the horizontal scale of the drawing is made one-half the vertical scale. The flow net is then drawn in the usual way and the rate of flow obtained from this diagram is true for the original construction for soil with a permeability $\sqrt{(k_x k_z)}$ in both horizontal and vertical directions.

A drawing of the flow net to a natural scale is shown in Fig. 4.9b, in which the vertical scale is unaltered, but the horizontal lengths are increased to their original value. The flow net so altered represents the actual equipotential and flow lines for the anisotropic soil, but the two sets of lines no longer intersect at right angles.

The validity of this construction can be shown thus. Fig. 4.9c shows an element of soil through which flow takes place. If the drop in head through the element in the horizontal direction is Δh_1 and that in the vertical direction is Δh_2, the total flow q through the element is given by

$$q = \Sigma A k i = b k \frac{\Delta h_1}{b} + b k \frac{\Delta h_2}{b}$$

$$= k(\Delta h_1 + \Delta h_2).$$

On the natural scale diagram the element takes the form shown in Fig. 4.8d. The flow is now

$$q = b k_x \frac{\Delta h_1}{b\sqrt{(k_x/k_z)}} + b k_z \sqrt{(k_x/k_z)} \frac{\Delta h_2}{b}$$

$$= \sqrt{(k_x k_z)} \, (\Delta h_1 + \Delta h_2).$$

These values of q are equal, therefore $k = \sqrt{(k_x k_z)}$.

4.4 QUICKSANDS AND PIPING

When seepage of water through soil is in an upward direction an unstable situation can arise when the upward seepage force induced by the flow balances or exceeds the downward force due to gravity.

Quicksand
When the seepage pressure due to an upward flow of water in sand exceeds the downward force of gravity the sand is in the state known as *quicksand*. Thus quicksand is not a special type of material, but a condition which any granular material may attain under certain circumstances.

Fig. 4.10 shows an example of how quicksand may be met with at the bottom of an excavation. The downward effective pressure at depth d is $(\gamma - \gamma_w)d$.

The head causing flow through the depth of sand d is h. The hydraulic gradient (i) causing flow is therefore h/d.

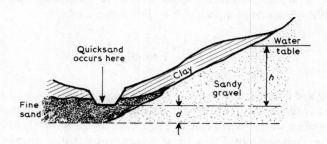

Fig. 4.10 Quicksand conditions.

Seepage force per unit volume = $i\gamma_w = (h/d)\gamma_w$.

The seepage pressure is obtained by multiplying the seepage force per unit volume by the distance through which flow takes place.

Therefore seepage pressure = $(h/d)\gamma_w d = h\gamma_w$.

When the upward seepage pressure, which is exerted as an upward drag on the particles of sand, is equal to the downward effective pressure, a critical condition ensues. If the seepage pressure increases beyond this critical point, the particles will be suspended and quicksand conditions occur. This critical state is represented by

seepage pressure = static effective pressure

$$h\gamma_w = (\gamma - \gamma_w)d$$

or

$$\frac{h}{d} = \frac{\gamma - \gamma_w}{\gamma_w}$$

which is known as the *critical hydraulic gradient*.

Replacing γ by $[(G_s + e)/(1 + e)]\gamma_w$ the expression becomes

$$\text{critical hydraulic gradient} = \frac{G_s - 1}{1 + e}.$$

Theoretically, therefore, the size of the sand grains has no effect, though it is well known that quicksand is of more frequent occurrence in fine sands. Fine sands are usually of fairly uniform grain-size and in a loose state of packing, and therefore have a relatively high void ratio and a correspondingly low critical hydraulic gradient.

The critical hydraulic gradient usually has a value of approximately unity. It would be exactly 1 for a sand for which $G_s = 2.65$ and $e = 0.65$ – both fairly average values. For such a sand

$$\frac{G_s - 1}{1 + e} = \frac{2.65 - 1}{1 + 0.65} = 1.$$

Piping and internal erosion

Piping is the term applied to an unstable condition which sometimes arises when water seeps through a soil mass. As the water flows in the manner depicted by the flow net it experiences a gradual drop in head as shown by the spacing of the equipotential lines. This drop in head is accompanied by a seepage force which represents the transfer to the soil particles of the head lost as the water passes through the soil. If this seepage force acts in the direction of gravity it increases the apparent weight of the soil particles and makes them bed more closely together. If it acts upwards against gravity, it decreases the weight of the particles, and the shearing strength and stability of the soil. When the upward drag on the particles becomes equal to their submerged weight, the particles can offer no frictional resistance and can be considered to be afloat. The sand and water under these circumstances act together as a liquid. The water carries with it the material of the foundation, a 'pipe' being formed through which internal erosion takes place. Piping of this kind leads ultimately to complete failure of the foundation.

It has been shown above that the seepage force per unit of volume on the particles is equal to the hydraulic gradient multiplied by the unit weight of water. The danger of piping occurs when the hydraulic gradient is high, that is when there is a rapid loss of head over a short distance. Such a condition is shown in the flow net by a close network of squares. If, therefore, in the flow net an area of small squares is observed in a region of upward flow, conditions there should be examined.

In designing cofferdams or other structures likely to be affected by seepage the stability of the soil against piping should be carefully checked. If the preliminary design does not show an adequate factor of safety against this type of failure, suitable modifications should be made, usually taking the form of lengthening the path of percolation. In a cofferdam this can be effected by increasing the depth. A remedial measure sometimes adopted when piping does occur is to apply a surcharge load on the downstream side extending for a width $d/2$ from the face, d being the depth of the cofferdam below downstream ground level. Care must be taken to provide a heavy filter of permeable material so that the hydraulic gradient is unaltered. The length of the seepage path under a dam can be lengthened by providing a sheet-piling cut-off under the base; another method is to construct an impervious apron for a short distance upstream from the base

4.5 APPLICATION OF THE WORK OF CHAPTER 4

Application 4A

Estimate the seepage loss through the dam shown in Fig. 4.7a if the head is 15 m and the permeability 5×10^{-6} m/s.

(i) In the flow net there are five channels and fourteen steps of potential. The flow through the dam, per metre of its length, is, therefore,

$$q = kH\tfrac{5}{14}.$$

(ii) Substituting the values from the problem:

$$q = 5 \times 10^{-6} \times 15 \times \tfrac{5}{14} = 26 \cdot 8 \times 10^{-6}\, \text{m}^3\, \text{s}^{-1}.$$

Application 4B

Find the seepage pressure at the point X beneath the dam shown in Fig. 4.5.

(i) The total head to be dissipated through the construction is H feet. At the point X the excess hydrostatic head is represented by the number of squares between X and the tailwater, namely 12.

(ii) The excess hydrostatic pressure at X is, therefore, the seepage pressure and is (for $H = 16$ m, and $\gamma_w = 10$ kN m^{-3})

$$12H\gamma_w/15 = 12 \times 16 \times 10/15 = 128\, \text{kN m}^{-2}.$$

Similarly, at the point Y the seepage pressure is

$$11H\gamma_w/15 = 11 \times 16 \times 10/15 = 117\, \text{kN m}^{-2}.$$

Application 4C
The soil shown in Fig. 4.11 is subjected to a head of 13 m. Find the factor of safety against piping.

(i) In the region of point *P* the flow is upwards and there is a rapid loss of head towards the surface. Therzaghi and Peck have shown that the critical zone in which the piping is liable to take place is over a width in front of the piles of about half the depth of penetration, as shown by *TU* in Fig. 4.11. In the example the mass of soil is a prism 10 m deep x 5 m wide x 1 m thick. The seepage force on this mass of soil corresponds with the total loss of head from the base of the prism to the surface, since, at the surface of the soil towards which the water is flowing, the head is zero.

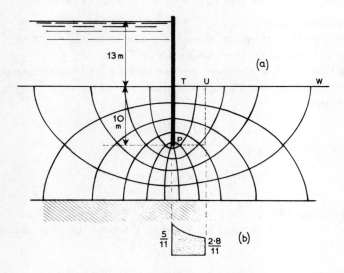

Fig. 4.11 Conditions for piping (Application 4C).

(ii) The pressure over the base of this prism is found by observing where the equipotential lines cut the base. From the toe of the pile to the sand surface *TW* there are 5 drops of head out of the 11 which represent the total head lost, and at the right-hand edge of the prism there are 2·8 drops of head. Fig. 4.11.b shows how the pressure varies across the bottom of the prism. The mean pressure is represented by about 3·6 drops of head or

$$\frac{3·6}{11} \times 13 \text{ m} = 4·25 \text{ m of lead.}$$

(iii) The seepage force can be found in two ways. It has been shown that the seepage force per unit volume is equal to the hydraulic gradient multiplied by the weight density of water, that is, $i\gamma_w$ kN m^{-3}. The head of 4.25 m is lost in the passage of the water from the level of the point of the pile to the surface (10 m), and the mean hydraulic gradient is thus

$$\frac{4·25}{10} = 0·425.$$

The seepage force per unit volume is, therefore,

$$0·425 \times 9·81 \text{ kN m}^{-3}$$

and, since the volume of soil concerned is

$$10 \times 5 \times 1 = 50 \text{ m}^3$$

the total seepage force is

$$0·425 \times 9·81 \times 50 = 208 \text{ kN.}$$

(iv) Another way of looking at this problem is to refer the seepage pressure to its original cause: loss of head. The head at P is 4·25 m of water

$$4·25 \times 9·81 \text{ kN m}^{-2}.$$

This represents a total upward excess hydrostatic force, over the 5 m^2 concerned, of

$$4·25 \times 9·81 \times 5 = 208 \text{ kN.}$$

The same result is thus obtained as when considering seepage pressure on the soil, for the loss of hydrostatic head is transferred to the soil as a 'drag'.

(v) The submerged weight of the soil (whose bulk density is 1·8 Mg m^{-3} is $(1·8 - 1)$ 9·81 kN m^{-3}, and the total weight is, therefore,

$$0·8 \times 9·81 \times 10 \times 5 = 392 \text{ kN.}$$

(vi) The factor of safety against piping is represented by the

67

weight of the soil divided by the seepage force

$$\frac{392}{208} = 1 \cdot 9$$

(vii) If it is desired to raise the factor of safety to, say, 2·2, the downward weight of soil and filter must be

$$2 \cdot 2 \times 208 = 457 \text{ kN}.$$

The extra load required in the form of a permeable filter is

$$457 - 392 = 65 \text{ kN}$$

or 13 kN m^{-2} over the 5 m^2 which represents the area likely to suffer piping.

CHAPTER 5

Shearing resistance

The *shear strength* of a soil is its maximum resistance to shearing stresses. When this resistance is exceeded failure occurs, usually taking the form of surfaces of slip. Shear strength is usually assumed to be made up of:

(i) *internal friction,* or the resistance due to interlocking of the particles;

(ii) *cohesion,* or the resistance due to the forces tending to hold the particles together in a solid mass.

Generally speaking, coarse-grained soils such as sands derive their shear strength almost entirely from intergranular friction, but with other soils the strength is a combination of both forms of resistance. It is therefore convenient to consider three conventional types of soil:

(i) coarse-grained, frictional, or cohesionless;

(ii) fine-grained or cohesive ($\phi = 0$);

(iii) cohesive-frictional or 'c-ϕ'.

Most clay soils, in time, derive part of their shearing resistance from internal friction, but, as will be explained later, when loaded under conditions of no change in moisture content, appear to possess cohesion only.

5.1 CONDITIONS OF SHEAR FAILURE

A law governing the shear failure of soils was first put forward in 1773

69

by Coulomb in the form

$$s = c + \sigma \tan \phi.$$

The relationship is shown diagrammatically in Fig. 5.1, in which s is plotted against σ.

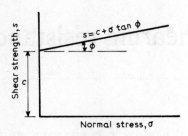

Fig. 5.1 Coulomb's law.

Effective normal stress

In Coulomb's equation as stated above it is incorrectly assumed that the strength of the soil is governed by the total normal stress σ on the shear plane. The application of stress usually results in a temporary increase in the pore-pressure, and the effective or intergranular stress σ', on which the shear strength depends, is given by

$$\sigma' = \sigma - u.$$

A more fundamental form of Coulomb's equation is therefore

$$s = \sigma' + (\sigma - u) \tan \phi' = c' + \sigma' \tan \phi'.$$

The symbols c' and ϕ' indicate that these constants refer to *effective stress* instead of *total stress*. If the void ratio, density, and pore-pressure remained constant during the process of shearing, c' would be the true cohesion and ϕ' the true angle of internal friction. For clays the true values are difficult to measure and even if known would be unsuitable for practical application.

The quantities c and ϕ referred to total stresses and c' and ϕ' referred to effective stresses, depend not only on the type of soil but also on the moisture content and on the conditions of testing or loading in the field. They should therefore be regarded as empirical constants and not as fundamental soil properties. The term c (or c') is usually called the *apparent cohesion*, while ϕ (or ϕ') is known as the *angle of shearing resistance*.

When pore-pressure measurements cannot be made it is common practice to determine the apparent cohesion and angle of shearing resistance with reference to total stress, the conditions of the test being chosen to represent as closely as possible those of the practical problem under consideration. However, the principle of effective stress is extensively applied to practical problems of stability analysis and design. For this purpose the shear parameters c' and ϕ' referred to effective stress are required.

Mohr Circle of stress

The conditions of shear failure are most conveniently depicted by Mohr's Circle of stress, which represents the state of stress at a specific point within a stressed material. It is assumed that the reader is familiar with this construction and its underlying principles.

Referring to Fig. 5.2a, the line *AB*, often called the *Coulomb Line*, or *failure line*, represents the conditions of shear failure in accordance with Coulomb's law. The Mohr Circle shown, touching this line at *R*, represents a condition of incipient failure. Any circle falling below and clear of the line *AB* would denote a condition of safe stress.

If the direction of the axis *OX* corresponds to the plane on which the major principal stress σ_1 acts, the line $P_3 R$ gives the direction of the plane of rupture, that is, the plane on which shear failure takes place. The co-ordinates of the point *R* give the normal stress σ and the shear stress s on this plane.

It can easily be shown from the geometry of the figure that the inclination of the plane of rupture to the line *OX* is $45° + \phi/2$ and that the shear stress on this plane is $\frac{1}{2}(\sigma_1 - \sigma_3)/\cos \phi$. It follows that, taking the Coulomb Line as the limit of safe stress, the maximum shear stress depends on the difference of the greatest and least principal stresses. This difference, $\sigma_1 - \sigma_3$, is commonly known as the *deviator* stress.

The condition of failure thus defined is independent of the intermediate principal stress σ_2. In the triaxial test described later, in which an all-round lateral pressure is applied, $\sigma_2 = \sigma_3$, but in practical problems these two principal stresses are not always equal. The effect of the intermediate principal stress on the conditions of failure has been the subject of research.

Fig. 5.2b represents the case of cohesive soil when $\phi = 0$. All Mohr

Circles touching *AB* and so representing incipient failure have the same radius, that is $s = c$. The plane of rupture P_3R makes an angle of 45° with the axis.

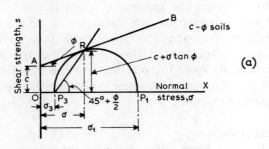

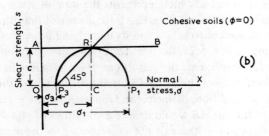

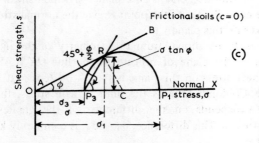

Fig. 5.2 Conditions of failure.

In coarse-grained soils such as dry sand or gravel, cohesion is negligible and the Mohr Circle diagram takes the form shown in Fig. 5.2c. In this case the ratio of the principal stresses σ_1/σ_3 is the same for all circles touching the line *AB*.

72

From Fig. 5.2c,

$$\sin \rho = \frac{RC}{OC} = \frac{\frac{1}{2}(\text{difference of principal stresses})}{\frac{1}{2}(\text{sum of principal stresses})} = \frac{\sigma_1 - \sigma_3}{\sigma_1 + \sigma_3}.$$

From this it is deduced that

$$\frac{\sigma_1}{\sigma_3} = \frac{1 + \sin \phi}{1 - \sin \phi}.$$

Under conditions of effective stress the apparent cohesion c' is often quite small or may even be zero, and the Mohr diagram approximates to the form of Fig. 5.2c.

5.2 MEASUREMENT OF SHEAR STRENGTH

For an ideal 'c-ϕ' material the coefficients c and ϕ are approximate constants which can be determined by carrying out two or more tests with different normal pressures acting on the plane of shear failure. If the shear strength is measured directly (as in the shear box test) the shear stress at failure is plotted against the stress normal to the shear plane as in Fig. 5.1 giving a straight line of which the slope is equal to $\tan \phi$ and the intercept on the axis of shear stress is equal to c. In the

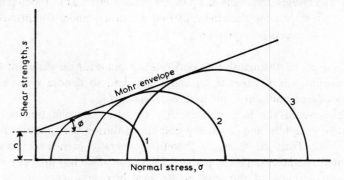

Fig. 5.3 Experimental determinations of c and ϕ.

triaxial test a cylindrical specimen is subjected to an all-round lateral pressure σ_3 and to a gradually increasing axial pressure, σ_1. By plotting σ_1 and σ_3 from the origin O, the Mohr Circle is obtained. When two or more of these circles have been drawn the common tangent (AB in Fig. 5.3), often called the *Mohr envelope,* determines the required values of c and ϕ (or of c' and ϕ' when effective stresses are plotted).

For soils, however, the coefficients c and ϕ vary with the moisture content and density, and thus depend not only upon the state of the soil before loading but also on the drainage conditions which apply during the process of loading and shearing. Also, for some conditions of testing the Mohr Envelope is only approximately straight.

Conditions of shear tests
The conditions of shearing tests for soils may be classified as follows:

(i) *Immediate or undrained tests*: the samples are subjected to an applied pressure under conditions in which drainage is prevented. The samples are then sheared, also under conditions of no drainage.

(ii) *Consolidated-undrained tests*: the samples are allowed to consolidate under an applied pressure, and then sheared under conditions of no drainage.

If the shear parameters c' and ϕ' referred to effective stress are required, measurements of pore-pressure are made during the test.

(iii) *Drained tests*: the samples are consolidated as in the previous test, but the shearing is carried out slowly under conditions of no excess pore-pressure.

In order to distinguish the coefficients c and ϕ for the different test conditions, the suffixes u, cu and d are used to denote undrained, consolidated-undrained, and drained, respectively.

Because of the high permeability of sand, consolidation occurs relatively rapidly and is usually completed during the application of the load. Tests on sand are therefore generally carried out under drained conditions. This is perhaps fortunate, since it is impossible to prevent drainage of the sand in the shear box, and hence undrained tests can be made only in the triaxial apparatus.

For soils other than sands, the choice of test conditions depends on the purpose for which the shear shrength is required. The guiding principle is that the drainage conditions of the test should conform as closely as possible to the conditions under which the soil will be stressed in practice.

Undrained tests are used for problems relating to saturated clay soils when changes in stress produce no immediate changes in moisture content and therefore no change in volume. Such problems include the stability of foundations, since during the period of construction very little consolidation will have taken place and consequently the change of moisture content will be negligible. For clay slopes also, undrained tests are often used for the investigation of short-term stability.

Consolidated-undrained tests are used where changes in moisture content are expected to take place, owing to consolidation, access of water or other causes, before the soil is fully loaded. In these tests, pore pressures are generally measured, giving the parameters c' and ϕ' referred to effective stress, which can then be used in problems in which the pore-pressures can be measured or estimated. An important example is the stability of earth dams during construction, when retaining water and under *rapid drawdown* conditions, that is when the water is lowered at a faster rate than that at which the material of the dam can consolidate.

As mentioned above, drained tests are nearly always used in problems relating to sandy soils. In clay soils, drained tests can be used for long-term stability problems, but consolidated-undrained tests with pore-pressure measurements are more usual.

5.3 SHEAR STRENGTH OF SAND

The development of shearing resistance of sands is best illustrated by reference to a typical stress/strain curve obtained from a shear box test (Fig. 5.4a). As the strain is gradually increased the shearing resistance builds up until the *peak stress* is reached and failure begins. As the strain is further increased the resistance falls to a steady value known as the *ultimate stress*. During this period of incipient failure, the arrangement of the soil particles is disturbed and the packing is loosened. The ultimate shearing resistance is therefore approximately the same as that of the sand in the loose state.

Assuming the Coulomb equation $s = \sigma \tan \phi$ to apply, two values of ϕ are obtained, one for the compact state, calculated from the peak stress, and the other for the loose state, calculated from the ultimate. The angle ϕ varies to some extent with the normal pressure on the plane of shear. Some clean sands exhibit slight cohesive strength under certain conditions of moisture content, owing to capillary tension in the water contained in the voids. The cohesion of clean sand, which is, in any case, very small, is thus variable with change in moisture content and should not be relied upon in strength calculations. On the other hand, even small percentages of silt and clay in a sand give it cohesive properties which may be sufficiently great to be taken into account in stability analysis.

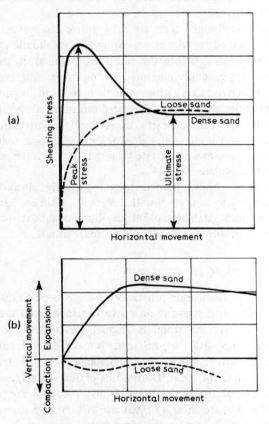

Fig. 5.4 Stress/deformation curves for sand.

The effect of shearing on dense and loose sand is apparent from Fig. 5.4b which shows the horizontal and vertical displacements during a shear box test. It will be seen that the dense sand expands until failure occurs, when the shearing resistance drops, and after this a slight decrease in volume may occur. Loose sand, on the other hand, compacts under the shearing action and its ultimate resistance approximates closely to that of the dense sand after failure. There is thus a *critical density,* or range of densities, for which the change of volume under the action of shear is negligible.

Thus sand in the loose condition, that is with its density less than the critical, is liable to 'shake down' under the action of vibrations or disturbances such as slight earth tremors or nearby pile-driving. The result is settlement, and if the sand is saturated its stability may be seriously impaired. Naturally occurring sands and gravels usually have a density greater than the critical and are therefore not affected in this way, but some deposits of fine uniform sand occur at densities below the critical and may in consequence be treacherous.

The critical density of sand can be determined by carrying out the shear box test at several different degrees of compaction. The critical density is that at which neither expansion nor contraction takes place during shearing.

5.4 SHEAR STRENGTH OF CLAY

The shear strength of saturated clay depends to a great extent upon its geological history. For normally consolidated clay the undrained shear strength is proportional to the pressure under which the soil has been consolidated (Fig. 5.5a). It follows that there are definite relations between shear strength and void ratio and between shear strength and moisture content, and that in uniform soil of this nature the undrained strength increases uniformly with the depth.

Effect of overconsolidation

In Fig. 5.5b, the full line is the pressure/void ratio curve for normally consolidated clay, often known as the *virgin consolidation curve.* If the compressive load is increased to a certain point, then released, the soil has been over-consolidated. If the load is afterwards reapplied, the void ratio follows the dotted curve.

The effect on the shear strength can be found thus. When the void ratio is e_0 the pressure is p_1, but since the corresponding pressure on the virgin curve is p_0 the shear strength will be that corresponding to the pressure p_0. A point on the pressure/shear strength curve for the overconsolidated clay can be obtained by projecting the point D at pressure p_0 horizontally to meet the ordinate at p_1 at D'. The complete curve for the overconsolidated clay is of the form ACB.

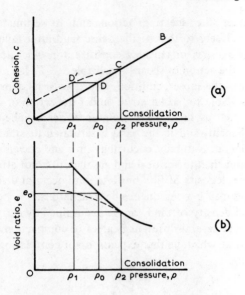

Fig. 5.5 Relations between cohesion, consolidation, pressure and void ratio.

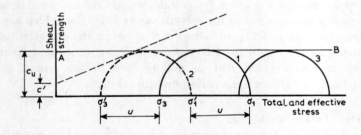

Fig. 5.6 Undrained tests on clay.

Drainage conditions

Undrained test. Undrained shear tests on saturated clay always show the angle of shearing resistance $\phi_u = 0$. This may be explained by considering an ideal material consisting of an incompressible soil skeleton with the voids filled with water. Under conditions of no drainage there can be no change in volume.

In Fig. 5.6 circle (1) represents the results of a triaxial test plotted on a base of total stress. The application of the cell pressure σ_3 is resisted by an increase of pore-pressure and the effective stress at failure remains the same. If u is the total pore-pressure the effective stress, circle (2) is obtained by shifting circle (1) a distance u along the axis, since $\sigma_3{}' = \sigma_3 - u$. Circle (3) is the total stress circle for a second test at the same moisture content, but with a different cell pressure. The corresponding effective stress circle must coincide with (2) and therefore circle (3) must be of the same diameter as (1). It follows that the Coulomb Line AB is horizontal and that $\varphi_u = 0$.

Consolidated-undrained test. Here the test is carried out in two stages. During the first stage drainage, and therefore consolidation, under a certain pressure p is allowed to take place. This pressure may be that of the existing overburden or some pressure which, it is intended, will be applied in the future. The initial consolidation can perhaps best be regarded as a form of specimen preparation. During consolidation the void ratio of the soil has been reduced from e_1 to e_2 with a corresponding increase in cohesion. The specimen is then sheared under undrained conditions. If several specimens consolidated under the pressure p_2 are sheared with different cell pressures σ_3, the results will be similar to those of an immediate undrained test. Thus $\phi_{cu} = 0$, but the apparent cohesion now has a value corresponding to the new void ratio.

Drained test. The specimen is first consolidated under a certain pressure and then sheared slowly under conditions of full drainage. Since the excess pore-pressures are dissipated the effective stress is equal to the total stress. For each test the shear strength corresponds to the void ratio at which failure takes place. From the envelope of the Mohr Circles, which is sometimes only approximately straight, the constants for the drained test, c_d and ϕ_d, are obtained.

Undrained tests with pore-pressure measurement. When the pore-pressure is measured during consolidated-undrained tests, it is possible to determine the Coulomb constants c' and ϕ' referred to effective stresses as given by the equation

$$s = c' + (\sigma - u) \tan \phi'.$$

The values of c' and ϕ' so found generally agree closely with c_d and ϕ' as determined from drained tests.

Residual shear strength of overconsolidated clay

Many instances have occurred of failures in overconsolidated clay slopes which had remained stable for many years previously. Analyses of such failures show that the shearing strength of the soil in the region of the failure is considerably lower than the usual strength as measured in the laboratory by conventional methods. The drop is strength is probably due to a strain-softening effect, leading to progressive failure.

Shearing tests, using the shear box, carried out with large strains, yield a stress/strain curve of the form shown in Fig. 5.7. The point of maximum stress is normally taken as the condition of failure, but as straining is continued the shear stress gradually falls off. At a strain of the order of 15 to 20 per cent it attains a steady value known as the *residual strength*.

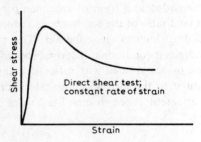

Fig. 5.7 Stress/deformation curve for overconsolidated clay.

From a series of large strain tests the residual shear stress s_r is found to be

$$s_r = c_r' + \sigma_r' \tan \phi_r'.$$

The cohesion intercept c_r is usually very small. If this is neglected the equation becomes

$$s_r = \sigma_r' \tan \phi_r'.$$

The residual shearing strength is of importance in circumstances where the soil has already suffered high strains. If a bank or cutting has slipped already, any calculation for remedial measures must use the residual rather than the peak shear strength. In over-consolidated, fissured clays especially, there is a gradual reduction in stability with time and increasing strain. In such materials the difference between the peak stress (usually recorded as the strength of the soil) and the residual shear strength, is more marked than in normally-consolidated clays. Thus, in the design of cuttings and embankments in over-consolidated clays, the residual strength should be determined for use in stability calculations.

Shear strength of cohesive soil when $\phi = 0$

The soils in this category include unsaturated soils as well as certain saturated soils, notably silts, which exhibit the property of dilatancy, also some saturated soils in which negative pressure may produce cavitation. In unsaturated soils, under test conditions of no drainage, some immediate volumetric change takes place because of compression of the air under the action of the applied pressure. The consequent reduction in void ratio produces an increase in cohesion which results in the inclination ϕ_u of the Mohr envelope not being zero. The subject of the strength of unsaturated soils, however, has not yet been fully investigated.

Shear strength of remoulded clay

The effect of remoulding on the compressibility of clay is discussed in a later chapter. Another important effect, known as *sensitivity,* is softening or reduction of shear strength, which is probably due to two causes: disturbance of the arrangement of the molecules in the adsorbed layers, and alteration to the original structure of the soil.

The *degree of sensitivity S* is defined as the ratio of the shear strength of the undisturbed soil to that of the remoulded material. The *remoulding loss* is the ratio of the reduction in strength to the undisturbed strength. For moderately sensitive clays the remoulding loss approximates to the liquidity index.

Sensitivity varies with the composition and structure of the clay, its consolidation history, and its liquidity index. Thus overconsolidated clays of low liquidity index show little remoulding loss and are termed unsensitive. Most ordinary clays, however, show a degree of sensitivity of 2 to 4. Clays for which S is from 4 to 8 are termed *sensitive clays,* while those for which S exceeds 8 are termed *extra-sensitive.*

That part of the loss of strength which is due to disturbance of the adsorbed layers is generally recovered if the clay is left with its moisture content unaltered. This phenomenon is known as *thixotropy*.

The sensitivity of clay is often of great practical importance, as for example in the bearing capacity of piles driven into clay. Again, an incipient slip in a clay bank may, if the clay is sensitive, develop into a serious slide, owing to the softening of the disturbed soil. The procuring of samples of sensitive clays for test purposes presents difficulties since even slight disturbance during sampling may have considerable effect on the strength.

Variation of shear strength with depth
In normally consolidated clay the shear strength shows an approximately uniform increase with depth. At any depth the clay has been consolidated by the pressure of the overburden, and, as will be seen by reference to Fig. 5.5a, the ratio c/p is the slope of the line *OB*. It has been found, however, that the rate of increase of shear strength with depth, that is, the ratio c/p, is often appreciably less than would be expected from consolidated-undrained tests on undisturbed samples. This discrepancy is probably due to the lateral pressures in the ground which have contributed to the natural consolidation being less in relation to the vertical pressures than in the laboratory consolidation process.

5.5 DEFORMATION OF SOIL DURING SHEAR FAILURE

In many problems the deformation produced by stress before actual failure occurs is of great importance. It has been shown that in the simplified case of undrained cohesive soil ($\phi = 0$) failure occurs when

$$\frac{\sigma_1 - \sigma_3}{2} = Cu$$

and that when $c = 0$ and $\phi > 0$ the limiting condition is

$$\sigma_1/\sigma_3 = \frac{1 + \sin \phi}{1 - \sin \phi}.$$

The strain or deformation produced during loading to failure varies with the way in which the stresses are applied.

In the normal triaxial test procedure – the lateral pressure,

$\sigma_2 = \sigma_3$, is maintained constant and σ_1 is gradually increased;

In a special form of triaxial test in which initially $\sigma_1 = \sigma_2 = \sigma_3$; the axial principal stress σ_1 remains constant while the lateral pressures σ_2 and σ_3 are progressively reduced.

The stress-ratio/strain curve in these two examples varies with the method by which the ultimate stress is reached. Thus when dealing with practical problems in which deformation is of primary importance a study of the *stress path* can be very useful.

Many other forms of stress path can be devised, and much research work is being carried out, using special types of apparatus, to gain information about this aspect of the stress-strain behaviour of soils.

The stress-strain relation for soil under compressive loading is discussed in the chapter on compressibility and consolidation.

APPLICATION OF THE WORK OF CHAPTER 5

Application 5A

Three soils are tested in a constant-rate-of-strain shear box. The cross-sectional area of the box is 2600 mm². The results obtained at failure were:

Soil	Test No.	Horizontal shearing force (kN)	Vertical loading (kN)
K	1	0·081	0·027
	2	0·085	0·040
	3	0·090	0·067
L	4	0·056	0·089
	5	0·083	0·133
	6	0·125	0·200
N	7	0·051	0·111
	8	0·052	0·178
	9	0·053	0·222

Plot the above results and obtain the apparent cohesion and the angle of shearing resistance. What is the probable soil type of each sample?

(i) The first step is to obtain both the shear stress and the normal stress in terms of $kN\,m^{-2}$. The loads given in the last two columns must be divided by the cross-sectional area of the shear box expressed in m². The results are:

Shear stress: 31·2, 32·6, 34·6; 21·5, 31·9, 48·1; 19·6, 20·0, 20·4.
Normal stress: 10·4, 15·4, 25·8; 34·2, 51·1, 76·9: 42·7, 68·5, 85·4.

(ii) Plot normal and shear stresses as in Fig. 5.2 (without the Mohr's Circles). Soil K is a c-φ soil with apparent cohesion of 28 kN m⁻² and a φ of between 13 and 14°. L is a cohesionless soil (Fig. 5.2c) with φ lying between 32 and 33°. Soil N is purely cohesive (Fig. 5.2b) with an apparent cohesion of 20 kN m⁻² and a zero value of φ.

Application 5B
A sample of soil is tested in triaxial compression. It is in the undrained condition and pore-pressures are measured. The values of pore-pressure and deviator stress are as follows (the cell pressure is 276 kN m⁻²).

Deviator stress (kN m⁻²)	Pore-pressure u (kN m⁻²)
0	165
362	217
528	203
616	159
661	117
676	86
683	69
679	62

Determine the values of the pore-pressure coefficients B and Ā.

(i) The value of B is the ratio of the increase in pore-pressure to the increase in applied pressure. When the cell pressure is first applied this is an increase from zero and represents $\Delta\sigma$. The pore-pressure also increases from zero before any principal stress is applied to the sample. This increase in pore-pressure is Δu.

Thus $B = \Delta u/\Delta\sigma = 165/276 = 0.6$, which shows that the soil is unsaturated. Otherwise B would have been unity or thereby.

(ii) To obtain A for the various values of applied stress and pore-pressure it is necessary to obtain the values of the increases in pore-pressure for each increment of deviator stress. Increases (and decreases) from 165 kN m^2 are:

$$0, \; 52, \; 38, \; 6, \; 48, \; 79, \; 96, \; 103 \text{ kN m}^{-2}$$

for the eight values of deviator stress listed.

(iii) The value of the parameter or coefficient $\bar{A}$ is the product of the original coefficients B and A. The equation given in Section 3.2 for B and A becomes

$$\Delta u = B \, \Delta\sigma_3 + \bar{A}(\sigma_1 - \sigma_3).$$

Since σ_3 is kept constant, $\Delta\sigma_3$, is zero and the value of $\bar{A}$ is thus

$$\bar{A} = \frac{\text{change in pore-pressure}}{\text{deviator stress}} = \frac{\Delta u}{(\sigma_1 - \sigma_3)}.$$

(iv) Using the values of deviator stress and those of the change in pore-pressure as shown in (ii), the values of A are obtained by simple division, and result in the list:

$$+0.144, \; +0.072, \; -0.010, \; -0.073, \; -0.117, \; -0.141, \; -0.152$$

which are dimensionless.

Application 5C

A saturated sand sample (38 mm diameter, 76 mm long), was tested in triaxial compression. The cell pressure was 410 kN m^{-2}; the sample had a mass of 175 g; both change in length and change in volume were measured under increasing load. The readings obtained were:

Load applied by proving ring (kN)	Change in length of specimen (mm)	Change in volume of specimen (ml)
0	0	0
0·73	2·5	1·2
0·94	5·0	2·5
1·05	7·5	3·0
1·11	10·0	3·3
1·15	12·5	3·2

The test was carried out under drained conditions. Find the angle of shearing resistance under these conditions at failure.

(i) Read Section 5.2 and the discussion on the triaxial compression test in Chapter 14.

(ii) The mention of drained conditions means that the pore-pressure is dissipated throughout the test, so that no excess pore-pressure is set up during the test.

(iii) To obtain the true stress on the circular cross-section of the sample, it is necessary to calculate the average cross-sectional area of the sample at each increment of load. Using A for cross-sectional area, V for volume and L for length, together with subscripts o for 'original' and f for 'final' (i.e. under each increment of load), we have the following equation:

Final volume under any = original volume less
increment of load   change in volume

Final area x final length = original area x original length
                                       less change in volume

$$A_f L_f = A_o L_o - \Delta V$$
$$A_f(L_o - \Delta L) = A_o L_o - V_o \frac{\Delta V}{V_o}$$
$$A_f L_o - L_o \frac{\Delta L}{L_o} = A_o L_o - L_o \frac{\Delta V}{V_o}$$
$$A_f = A_o \frac{1 - \frac{\Delta V}{V_o}}{1 - \frac{\Delta V}{L_o}}$$

This gives the values of the final (mean) cross-sectional areas for each increment of load in terms of the measured parameters.

(iv) The calculation can now be set up in order to obtain the values of the final cross-sectional areas (see Table 5.1).

Table 5.1 *Deviator stress in a triaxial compression test*

Load from proving ring (kN)	$\dfrac{\Delta L}{L_o}$	$\dfrac{\Delta V}{V_o}$	$\dfrac{1 - \Delta L}{L_o}$	$1 - \dfrac{\Delta V}{V_o}$	A_f (mm^2)	Deviator stress (kN m^{-2})
0	0	0	1·00	1·00	1134	0
0·73	0·033	0·014	0·967	0·986	1157	631
0·94	0·066	0·029	0·934	0·971	1179	797
1·05	0·099	0·035	0·901	0·965	1213	866
1·11	0·132	0·038	0·868	0·962	1259	882
1·15	0·164	0·037	0·836	0·963	1304	882

$A_o = 1134$ mm^2, $V_o = A_o L_o = 1134 \times 76 = 86184$ mm^3.
The deviator stress is found by dividing the applied load by the mean final area (A_f).

(v) The deviator stress does not increase beyond 882 kN m^{-2} which is clearly the failing stress (plot deviator stress against longitudinal strain to see the effect). σ_1 is thus 1292 kN m^{-2} since σ_3 is 410 kN m^{-2}. Draw the Mohr Circle and assess the angle of shearing resistance ϕ. It should be about 31°.

CHAPTER 6

Stresses in soil from external loading

In the design of the foundations for engineering structures of all kinds it is necessary to study the way in which the loads are transmitted to the soil and the resulting stress distribution within the soil itself.

The study of the physical failure of foundation materials and the planning and design necessary to avoid such failure therefore involve three important considerations:

(i) the distribution and intensity of pressure between the footing and the foundation material (*contact pressures*);

(ii) the intensity of normal and shearing stresses at various points within the mass of the foundation material (the values of vertical pressures at various depths are used in the study of consolidation, and the values of shear stress in the study of shear failures);

(iii) the mechanism of failure of the foundation material when over-loaded, having regard to its physical characteristics as found by the tests described later in this book.

6.1 CONTACT PRESSURE DISTRIBUTION

It would seem reasonable to assume that a uniform load applied to a homogeneous soil should show a uniform contact pressure between the foundation and the soil. If the footing is perfectly flexible – if it is composed, for example, of a thin membrane – then a uniformly applied load will exert a uniform pressure on any type of soil. This extreme case of an ideal situation is not encountered in practice.

As the foundation slab increases in rigidity, the contact pressure departs from a uniform distribution. From the theory advanced by Boussinesq, and infinitely rigid strip foundation supported by an elastic medium produces a concave distribution of contact pressure. The pressure is infinite at the edges and decreases towards the centre. In practice, of course, no soil can withstand an infinite pressure, and the theoretical distribution is disturbed. It is, however, still approximately parabolic, showing a low stress at the centre of the strip and a high stress at the edges.

Schultze analysed eleven case records from various countries, and found that eight of the rigid foundations studied showed concave contact pressure distributions. His conclusion is that the theoretical expressions developed to define the distribution of contact pressure agree closely with measured results.

Modulus of deformation

The *elastic* or *Young's modulus* of materials provides an invaluable guide to behaviour. In soils the modulus can be used to estimate the *immediate settlement*. Soils, and other constructional materials such as concrete, are not elastic, and the problem of defining a modulus in the terms usually employed, becomes complex. Sometimes the stress/strain curve (familiar in the study of steel, for example), shows a straight portion on first loading, thus indicating elastic behaviour. However, if the same sample is loaded and unloaded repeatedly, without approaching the failing load, the stress/strain line does not remain in the same position. It moves along the strain axis, indicating creep and permanent deformation, and it also changes slope, indicating a different modulus from that shown originally. To make interpretation more difficult, the line may well be considerably curved.

In testing for modulus in the laboratory, the type of compression test influences the value. Tests can be drained, consolidated-undrained or undrained. Restriction of the ends of the sample during the test can also have a greater effect on the value of the modulus than might be expected.

In addition to the method of testing samples in the laboratory, a modulus may be obtained from field loading tests *in situ*. This has the advantages that the soil is not disturbed, and that a larger volume of soil is affected, but there are difficulties even here in the task of attempting to obtain a unique figure for a modulus. The size of the plate used has an important effect on the value of the modulus obtained. Under certain defined conditions, *in situ* results may be quoted as a *modulus of subgrade reaction*.

However the tests are carried out, when a figure is quoted of a modulus relating stress to strain, much care must be taken in its interpretation and use. The figure given may refer to first loading or to the different slopes obtained after repeated loading. It may refer to the slope of a tangent to one of the curved lines (*tangent modulus,* or *initial tangent modulus,* if measured on the first curve) or to what is known as the *secant modulus* (the modulus of subgrade reaction is defined as a secant modulus). The secant modulus is obtained by choosing an arbitrary strain value, locating this point on the stress/strain curve chosen, and joining this point to the origin by a straight line. The slope of this line is the secant modulus. Each of these methods indicates a different value of modulus, ostensibly for the same material. The figure of '*E*' should never be accepted without an enquiry as to how it was obtained.

From a wide range of tests at the University of Strathclyde, Dr W. F. Anderson suggests that the best type of test for consistent results is that on a 100 mm diameter sample with lubricated end platens in the compression test. The sample should first be consolidated and the test then carried out in the undrained condition. The sample should be loaded and unloaded, and then on the second loading, the stress/strain relationship should be obtained and a defined modulus measured from the curve.

6.2 THEORETICAL INTENSITY OF STRESS IN A SOIL MASS

The estimation of the intensity of stress in soil under external loading has been made both by theoretical investigation and by experiment. Neither method can be said to lead to a solution which is generally applicable to all types of foundation. For instance, experimental tests must be carried out on particular types of material with specific properties and most tests have been carried out on granular materials without cohesion. Mathematical analyses, on the other hand, must, for the sake of simplicity, be based on assumptions often not valid in practice – of simplified loading conditions and of ideal material.

Theories have been worked out for the following simplified conditions of contact pressure:

(i)   concentrated point or line load;
(ii) uniformly loaded contact area;
(iii) area with straight-line variation of contact pressure.

It is shown above that the distribution of contact pressure is in practice more complex, but, as the depth below the footing increases, errors due to incorrectly assumed contact pressure distribution become dissipated. It is obvious, therefore, that the results obtained either by experiment or by theory must be interpreted in the light of experience.

Stress distribution under a concentrated load
The mathematical analysis of the intensity of stress in a foundation material is based on the theory published by Boussinesq in 1885. He investigated the stress in a semi-infinite, elastic, isotropic and homogeneous continuum loaded normally on its upper plane surface

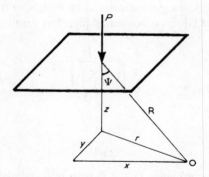

Fig. 6.1 Boussinesq co-ordinates.

by a concentrated load. Although no soils can be said to possess all the attributes indicated by this statement of the problem, yet many clays do at least approach the postulated conditions, and the Boussinesq figures form a valuable basis for the estimation of stress at some depth below the surface.

The intensities of the stresses at point O (Fig. 6.1), defined by coordinates x, y, and z, were determined by Boussinesq to be as follows:

Vertical direct stress on horizontal planes at depth z

$$= \frac{3Pz^3}{2\pi R^5} \quad \text{or} \quad \frac{3P}{2\pi z^2} \cos^5 \psi$$

Vertical shear stress $= \dfrac{3Prz^2}{2\pi R^5}$.

Horizontal direct stress (parallel to y-axis)

$$= \frac{P}{2\pi} \left[\frac{3y^2 z}{R^5} - (1 - 2\nu) \left(\frac{y^2 - x^2}{Rr^2(R + z)} + \frac{x^2 z}{R^3 r^2} \right) \right]$$

where ν = Poisson's Ratio.

These expressions are independent of Young's Modulus, but the horizontal stress does involve Poisson's Ratio.

For estimating the consolidation of foundation material, the vertical pressure on horizontal planes is of primary importance. The first of the expressions given above can be written in the form:

Vertical direct stress on horizontal planes at depth $z = K(P/z^2)$

where

$$K = \frac{3}{2\pi} \frac{1}{\left[1 + \left(\dfrac{r}{z} \right)^2 \right]^{5/2}}$$

or

$$\frac{3z^5}{2\pi R^5}$$

Values of the influence factor K have been tabulated over a wide range of the ratio r/z or $\tan \psi$ and are shown in Table 6.1 at the end of this chapter. Geedes (*Geotechnique,* 1966 and 1969) has further elaborated these figures.

In practice the load imposed by a footing is spread over a finite area. As the foundation material is assumed to be elastic and isotropic, the principle of superposition may be applied. The vertical stress under a finite loaded area can thus be found by integration from the Boussinesq equations. Alternatively, it can be calculated by dividing the loaded area into elements small enough to allow their loadings to be considered concentrated and summing up the effects of each loaded element.

Vertical pressure under a circular loaded area

The problem of determining the pressure under the centre of a uniformly loaded circular area arises in the design of foundations. By integration it is found that the vertical normal pressure at depth z below the centre of a circular area of diameter D carrying a load of q per unit area is qK where the influence factor

$$K = 1 - \frac{1}{\left[1 + \left(\dfrac{D}{2z}\right)^2\right]^{3/2}}.$$

Table 6.2, at the end of this chapter, gives the values of influence factors from which such pressures can be determined.

Stresses under a strip load

A rectangular footing which is long in comparison with its width, such as that of a wall, is most conveniently dealt with as a two-dimensional problem. The theoretical case of a concentrated load applied at the edge of a semi-infinite, elastic, and isotropic lamina can be readily solved by the theory of elasticity. From this solution the stresses produced by a strip footing, where the load is spread over a finite width, are obtainable by integration.

It can be shown that the principal stresses under a strip footing are given by

$$\text{maximum principal stress} = \frac{q}{\pi}(\beta + \sin \beta)$$

$$\text{minimum principal stress} = \frac{q}{\pi}(\beta - \sin \beta)$$

where β is the angle defined by Fig. 6.2 and q is the uniform intensity of loading. These expressions are independent of the elastic constants.

Although the values of the horizontal and vertical shearing stress at any given point in a foundation are easily determined by the use of the Boussinesq equations and the methods described above, yet the determination of the direction and intensity of the *maximum* shearing stress is a complex problem when the work is carried out in three

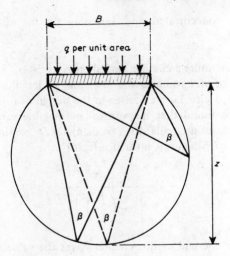

Fig. 6.2 Strip loading co-ordinates.

dimensions. It is usual, therefore, when determining the maximum shearing stress (and frequently other stresses in the foundation) to abandon the three-dimensional methods and to assume a strip load. This simplifies the problem by reducing it to a two-dimensional investigation. A section may be cut through loading and foundation at right angles to its length at a point remote from the ends, and only the x and z co-ordinates considered. The expressions for a strip footing can then be used to evaluate the stresses. The maximum shear stress is half the difference of the maximum and minimum principal stresses, and is equal to

$$\frac{q}{\pi} \sin \beta$$

which is constant so long as β is constant. In other words, the locus of the points at which the intensity of maximum shearing stress is constant is a circular arc, such as is shown in Fig. 6.3.

The greatest value which the maximum shear can have is q/π, or $0.318q$, which corresponds to a value of $\beta = 90°$. Fig. 6.3 gives lines of isoshear for a strip footing uniformly loaded, and Table 6.3, at the end of the chapter, gives influence factors for both shear stress and vertical pressure. Expressions for the various stresses in strip foundations under non-uniform loading have also been published.

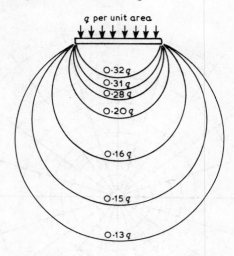

Fig. 6.3 Lines of isoshear for strip load.

Vertical pressure under a rectangular loaded area

One method of finding the vertical pressure at a point below a rectangular loaded area is to divide the area into a number of small squares or rectangles and to consider the load on each concentrated at its centre. The stresses at any point within the mass of soil under the rectangle induced by each of these concentrated loads are then added up to give the total stress at that point. The error involved in such an approximation is small, being less than 3 per cent if the length of the side of the small areas is less than one-third of the distance to the point O and within 2 per cent if the elements measure less than one-quarter of that distance.

Another method is to use influence factors, by means of which the pressure at any depth vertically below the *corner* of a uniformly loaded rectangular area can be calculated. These influence factors are shown in Table 6.4 (at the end of the chapter).

Let L be the length and B the breadth of the loaded area, and let z be the depth at which the vertical pressure is required. Using the appropriate values of L/z and B/z, the influence factor is read off from the table. The vertical pressure at this depth is then obtained by multiplying the influence factor by the contact pressure.

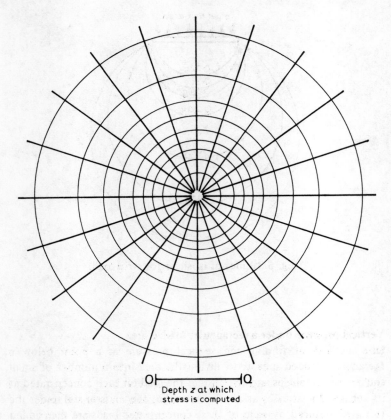

Depth z at which
stress is computed

Fig. 6.4 Influence chart for vertical pressure.

Influence charts
If a unit influence load is placed at various points on a beam,
influence diagrams can be drawn for quantities such as bending
moment or shearing force at some given section. In a similar manner,
an influence load placed at various positions on the surface of the
foundation can be used to determine the influence diagram for
intensity of stress at some point at a given depth. If a plan of the
loaded area to be studied is laid on the influence chart, it is possible to
determine the resulting intensity of stress for any point at the depth
for which the diagram is drawn.

Influence charts for all depths are similar to each other, but differ in scale. It is therefore convenient to draw one influence chart and vary the scale of the loaded area for each depth to be studied. Fig. 6.4 shows the influence chart devised by Newmark for vertical pressure at depth z (University of Illinois, U.S.A., Bulletin 338).

Each area on the chart, bounded by two arcs and two radii, has an influence value of 0·002.

The length OQ below the chart represents, to scale, the depth z at which the vertical pressure is required.

Using the scale to which OQ represents this depth z, the plan of the loaded area is superimposed on the chart. Then the required pressure is 0·002 times the number of influence areas within the loaded area, multiplied by the applied pressure.

Pressure bulb

The general shape of the lines of equal vertical pressure on horizontal planes beneath a loaded area has suggested the use of the term *pressure bulb* to designate the region of foundation material affected by the load. Figs 6.5a and b show the pressure bulbs for a small and a

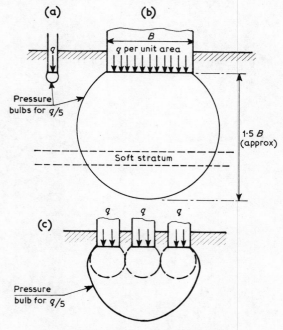

Fig. 6.5 Bulbs of pressure.

large loaded area. The outline connects points at which the vertical pressure is one-fifth of the intensity of the applied load. For a uniformly loaded square or circular area bearing on an elastic medium the depth of this bulb is 1·5 times the diameter or width of the footing. The pressure bulb for three small footings in Fig. 6/5a shows the cumulative effect of closely spaced loads.

The conception of the pressure bulb is of value in deciding to what depth investigations of soil properties should be carried. For example, a soft stratum which crosses the pressure bulb might have considerable effect on the stability of a large footing such as that in Fig 6.5b if the soft stratum lies well below the bulb, as in Fig. 6.5a, its influence must be very small.

It should be remembered that these analyses refer to loading on the surface. For sub-surface loadings, a closer estimate of theoretical stress is obtained from the tables derived by Geddes (*Geotechnique*). These give the stresses due to a force applied at a point below the surface of a semi-infinite solid.

6.3 TABLES OF INFLUENCE FACTORS

Table 6.1 *Influence factors for vertical pressure under a point load*

$\dfrac{r}{z}$	Influence Factor K	$\dfrac{r}{z}$	Influence Factor K	$\dfrac{r}{z}$	Influence Factor K
0·00	0·4775	1·00	0·0844	2·00	0·0085
0·10	0·4657	1·10	0·0658	2·10	0·0070
0·20	0·4329	1·20	0·0513	2·20	0·0058
0·30	0·3849	1·30	0·0402	2·30	0·0048
0·40	0·3294	1·40	0·0317	2·40	0·0040
0·50	0·2733	1·50	0·0251	2·50	0·0034
0·60	0·2214	1·60	0·0200	2·60	0·0029
0·70	0·1762	1·70	0·0160	2·70	0·0024
0·80	0·1386	1·80	0·0129	2·80	0·0021
0·90	0·1083	1·90	0·0105	2·90	0·0018

To obtain the vertical pressure at a given depth (point O in Fig. 6.2), multiply the appropriate influence factor by the value of the point load P and divide by z^2.

The vertical shear stresses are obtained by multiplying the values of vertical pressure by the appropriate r/z.

Table 6.2 *Influence factors for vertical pressure under centre of uniformly loaded circular area of diameter D*

$\dfrac{D}{z}$	Influence Factor K	$\dfrac{D}{z}$	Influence Factor K	$\dfrac{D}{z}$	Influence Factor K
0·00	0·0000	2·00	0·6465	4·00	0·9106
0·20	0·0148	2·20	0·6956	6·00	0·9684
0·40	0·0571	2·40	0·7376	8·00	0·9857
0·60	0·1213	2·60	0·7733	10·00	0·9925
0·80	0·1996	2·80	0·8036	12·00	0·9956
1·00	0·2845	3·00	0·8293	14·00	0·9972
1·20	0·3695	3·20	0·8511	16·00	0·9981
1·40	0·4502	3·40	0·8697	20·00	0·9990
1·60	0·5239	3·60	0·8855	40·00	0·9999
1·80	0·5893	3·80	0·8990	200·00	1·0000

To obtain the vertical pressure at a given depth under the centre of the circular loaded area, multiply the appropriate influence factor by the contact pressure q.

99

Table 6.3 *Influence factors for stresses under centre of uniformly loaded strip of width B*

$\dfrac{B}{z}$	β	Influence factor for shear stress $\dfrac{\sin \beta}{\pi}$	Influence factor for vertical pressure $\dfrac{1}{\pi}(\beta + \sin \beta)$
0·0	0° 0′	0·000	0·000
0·1	5° 44′	0·032	0·064
0·2	11° 26′	0·063	0·127
0·5	28° 04′	0·150	0·306
0·8	43° 36′	0·220	0·462
1·0	53° 08′	0·255	0·550
1·2	61° 56′	0·280	0·624
1·5	73° 44′	0·306	0·716
2·0	90° 00′	0·318	0·817
2·2	102° 40′	0·311	0·881
3·0	112° 38′	0·294	0·920
3·5	120° 30′	0·274	0·943
4·0	126° 52′	0·255	0·960
4·5	132° 04′	0·236	0·970
5·0	136° 14′	0·220	0·977
5·5	140° 02′	0·205	0·983
6·0	143° 08′	0·191	0·986
6·5	145° 44′	0·179	0·989
7·0	148° 06′	0·168	0·991
8·0	151° 56′	0·150	0·994
9·0	154° 56′	0·135	0·996
10·0	157° 24′	0·122	0·997
100·0	177° 44′	0·012	1·000
∞	180° 00′	0·000	1·000

To obtain the shear stress or the vertical pressure at a given depth, multiply the appropriate influence factor by the contact pressure q.

Table 6.4 *Influence factors for vertical pressure under the corner of a uniformly loaded rectangular area*

$\dfrac{Breadth}{Depth} = \dfrac{B}{z}$	\multicolumn{12}{c}{Length of rectangular area $= \dfrac{L}{z}$ Depth to stressed point}
	0·1	0·2	0·3	0·4	0·5	0·6	0·8	1·0	1·6	2·0	3·0	4·0
0·1	0·0047	0·0092	0·0132	0·0168	0·0198	0·0222	0·0258	0·0279	0·0306	0·0311	0·0315	0·0316
0·2	0·0092	0·0179	0·0259	0·0328	0·0387	0·0435	0·0504	0·0547	0·0599	0·0610	0·0618	0·0619
0·3	0·0132	0·0259	0·0374	0·0474	0·0559	0·0629	0·0731	0·0794	0·0871	0·0887	0·0898	0·0901
0·4	0·0168	0·0328	0·0474	0·0602	0·0711	0·0801	0·0931	0·1013	0·1114	0·1134	0·1150	0·1153
0·5	0·0198	0·0387	0·0559	0·0711	0·0840	0·0947	0·1103	0·1201	0·1324	0·1350	0·1368	0·1372
0·6	0·0222	0·0435	0·0629	0·0801	0·0947	0·1069	0·1247	0·1361	0·1503	0·1533	0·1555	0·1560
0·8	0·0258	0·0504	0·0731	0·0931	0·1104	0·1247	0·1461	0·1598	0·1774	0·1812	0·1841	0·1847
1·0	0·0279	0·0547	0·0794	0·1013	0·1202	0·1361	0·1598	0·1752	0·1955	0·1999	0·2034	0·2042
1·6	0·0306	0·0599	0·0871	0·1114	0·1324	0·1503	0·1774	0·1955	0·2203	0·2261	0·2309	0·2320
2·0	0·0311	0·0610	0·0887	0·1134	0·1350	0·1533	0·1812	0·1999	0·2261	0·2325	0·2378	0·2391
3·0	0·0315	0·0618	0·0898	0·1150	0·1368	0·1555	0·1841	0·2034	0·2309	0·2378	0·2439	0·2455
4·0	0·0316	0·0619	0·0901	0·1153	0·1372	0·1560	0·1847	0·2042	0·2320	0·2391	0·2455	0·2473

To obtain the vertical pressure at a given depth z under the corner of a rectangular loaded area $L \times B$, multiply the appropriate influence factor by the contact pressure q.

6.4 APPLICATION OF THE WORK OF CHAPTER 6

Below some depth, which differs for every structure and for its type of loading, the extra stress imposed by the structure is of no significance. Above that depth, the stresses imposed influence the consolidation and settlement experience by the underlying strata. The following applications illustrate the methods of obtaining the value of such stresses.

Application 6A

Find the vertical stress exerted in the soil vertically under a point P (Fig. 6.6) at a depth of 50 m. The uniform pressure over the surface of the raft is 300 kN m^{-2} .

(i) *Solution by influence factors from Table 6.4.* Divide the area into rectangles each of which has a corner above *P*, the point where the pressure is required. This is not always possible, when another method of solution must be used.

Table 6.5 *Influence factors for Application 6A*

Area No.			z = 50 m			
	L	B	Area (m^2)	L/z	B/z	Influence Factor
1	40	20	800	0·8	0·4	0·0931
2	20	20	400	0·4	0·4	0·0602
3	40	30	1200	0·8	0·6	0·1247
4	40	30	1200	0·8	0·6	0·1247
						0·4027

Multiplying the sum of the influence factors by the applied pressure, the stress at a depth of 50 m vertically under P is

$$0·4027 \times 300 = 120·8 \text{ kN m}^{-2}.$$

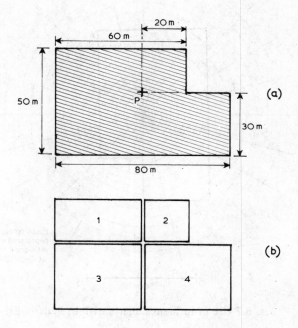

Fig. 6.6 Subdivision of irregular raft.

(ii) *Solution by Newmark's influence chart.* Fig. 6.7 shows the raft superimposed on the influence chart with the line *OQ* representing the depth of 50 m, to the scale of the plan. The point *P* is the centre of the chart. In practice the plan would be drawn on tracing paper and superimposed on a large-scale chart. The number of areas enclosed by the raft, including all the parts of area cut by the edges of the raft, is 202·4, by estimation.

The stress is, then: number of areas x bearing or contact pressure x value of one area as an influence factor.

$$\text{Stress at 50 m below } P = 202\cdot4 \times 0\cdot002 \times 300$$
$$= 121\cdot4 \text{ kN m}^{-2}.$$

103

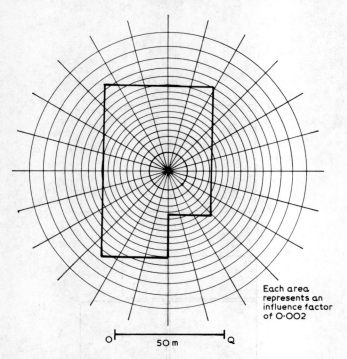

Each area
represents an
influence factor
of 0·002

O |⸻⸻⸻⸻⸻⸻⸻| Q
        50 m

Fig. 6.7 Use of influence charts with irregular area.

Application 6B

It is required to investigate the intensity of vertical pressure at 10 m below the centre of a square raft resting on clay and loaded with a uniformly distributed load of 250 kN m^{-2}. The raft is 8 m square.

(i) *Solution by Boussinesq's influence factors.* Table 6.1 gives values of Boussinesq influence factors according to the radius and depth of the recording point from the point of application of the point load. The uniformly distributed load can be considered as a number of concentrated point loads, if the raft area is divided up into small areas. As an example of the procedure, the number of areas chosen is 16 (Fig. 6.8). This could well have been more to obtain a more accurate grid.

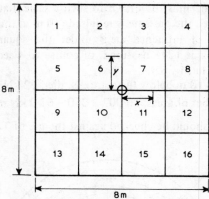

Fig. 6.8 Subdivision of loaded area.

Table 6.6 *Boussinesq influence factors (Application 6B)*

Square	x(m)	y(m)	r(m)	$\dfrac{r}{z} = \dfrac{r}{10}$	K
3	1	3	$\sqrt{10}$	0·32	0·374
4	3	3	$\sqrt{18}$	0·41	0·324
7	1	1	$\sqrt{2}$	0·14	0·455
8	3	1	$\sqrt{10}$	0·32	0·374
				Total	1·527

$\Sigma K = 4 \times 1 \cdot 527 = 6 \cdot 11.$

Only one-quarter of the raft need be studied, as the total pressure is four times the pressure caused by the quarter, in view of symmetry. Table 6.6 shows how the various required dimensions are used.

The value of K for the quarter area has been multiplied by 4 to include the whole raft.

Pressure at 10 m below centre of raft = (I.F. x load/z^2) (see Table 6.1).

The load on each of the 16 squares is 250 kN m^{-2} x 4 m^2 = 1000 kN.

Required pressure is 6·11 x 1000/100 = 61·1 kN m^{-2}.

(ii) *Solution by Newmark's influence chart.* Since the pressure is required below the centre of the raft, the centre is placed over the centre of the chart (or vice versa, if the chart is drawn on tracing paper). The square must be drawn to such a scale that OQ represents the depth at which the pressure is required (10 m in this problem).

The number of influence areas under the square, taking into account fractions, is 124, measured, of course to a scale much larger than that shown in Fig. 6.9.

Vertical stress 10 m below the centre is, then 62 kN m^{-2}:

Number of areas x 0·002 x 250 = 62·0 kN m^{-2}.

This solution should be checked against the use of Table 6.4.

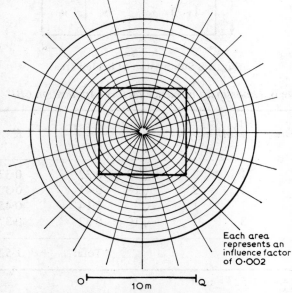

Each area
represents an
influence factor
of 0·002

o |⎯⎯⎯⎯⎯⎯⎯⎯⎯| Q
10 m

     Fig. 6.9 Use of influence chart with square footing.

CHAPTER 7

Pressures applied by a mass of soil

In the preceding chapter we have discussed the pressures set up in a mass of soil by the application of external loading. Next we have to consider the forces applied to a resistant structure, such as *earth pressure against a retaining wall*.

The two extremes of pressure to which a retaining wall may be subjected are:

(i) the *active pressure* of the soil resulting from a slight movement of the wall *away from* the filling, and

(ii) the *passive resistance* to slight displacement of the wall *towards* the filling.

The active pressure is a minimum value, and occurs when the wall is at the point of failure by moving away from the filling. The passive resistance is the maximum pressure to which the wall is subjected immediately before failure occurs by heaving up of the soil.

The magnitude of the active pressure and the effect such pressure has on the stresses within the wall have been the subject of much study and experiment. Two principal methods of approach have been adopted, generally known as Rankine's theory and the Wedge theory. Both depend on the conception that the soil filling is slightly expanded in a horizontal direction until surfaces of failure appear within its masses. In practice, such expansion of the material is achieved by a slight forward movement of a retaining wall – the

107

commonest method of failure – or by its rotation away from the filling, about an axis in its lower edge.

The allied subject of the passive resistance offered by soil to any slight movement of the retaining wall towards the filling is also of importance to civil engineers, and may be studied by methods similar to those adopted for the discussion of active pressure. The soil is considered to suffer a compression of an amount sufficient to cause surfaces of failure to appear within its mass.

Further variables which must be introduced into this subject are represented by the properties of the soils themselves The earliest derivations of earth pressure were made on the assumption that the soil was cohesionless. It was long before any successful attempt was made to understand the nature of the pressure exerted by soils possessing cohesion.

In this chapter the active pressure of both cohesionless and cohesive soils is first considered. This is followed by a study of passive r istance, and then attention is paid to the problems encountered when the surface supporting the soil fails by some movement other than pure translation or rotation about its lower edge.

The *coefficient of active earth pressure, K_a*, is defined as the ratio of the active earth pressure normal to a plane surface to the corresponding pressure in a fluid of the same density.

The *coefficient of passive earth pressure, K_p*, relates the passive earth pressure on a plane to the corresponding fluid pressure.

Earth pressure at rest. Although the active and passive pressures are of paramount importance in engineering design they do not represent the most usual pressure condition to which retaining walls are subjected. The active pressure which finally causes overturning or sliding of the wall occurs when the wall shows a tendency to movement. As the wall moves even slightly away from the filling, the pressure drops in value to that of the active condition. The earth pressure at rest, which the wall sustains without movement, is therefore, greater than the active earth pressure.

The *coefficient of earth pressure at rest, K_0*, is the ratio of the earth pressure on a plane to the corresponding fluid pressure, when there is no movement. Thus if the weight of soil above any depth z (the vertical pressure on a horizontal plane) is γz, then the horizontal earth pressure at rest is $K_0 \gamma z$.

Theoretical considerations indicate that for a wide range of normally consolidated soils, K_0 is approximately equal to $1 - \sin \phi'$, where ϕ' is the angle of shearing resistance under effective stress conditions. Experimental evidence, from laboratory tests and from field tests using pressure cells, indicate that this relation is reasonably correct under ideal conditions. In practice considerable variations may occur, from the structure of the soil mass, its past history, chemical changes, weathering, etc. For example, in overconsolidated clays K_0 has been increased by the removal of the overburden, but on reloading it is reduced.

Values of K_0 commonly adopted in practice are 0·4–0·6 for sands and 0·5–0·75 for clays. The coefficients of active (least) and passive (greatest) pressures are of greater importance to the engineer than the coefficient of earth pressure at rest.

7.1 ACTIVE PRESSURE OF COHESIONLESS SOIL

The two different methods of calculating earth pressure were developed mathematically by Coulomb in the eighteenth century and by Rankine in the nineteenth. Both made the assumption that the material behind the retaining wall is a dry cohesionless granular filling. It is seldom that such a filling (e.g. dry sand) is encountered in practice, but a study of Coulomb's and Rankine's original methods (both of which were later adapted and extended in scope) is of value in understanding the developments used at the present time.

Rankine's method when the surface is horizontal
Rankine did not study the pressure of the earth on the back of a retaining wall, but considered the stresses developed within the mass of soil by reason of its own weight. Fig. 7.1 shows that at a depth z an elemental portion of the soil is acted upon by two principal stresses; the maximum, γz, is equivalent to the weight of a column of soil of depth z, and the minimum, p_a, is the active pressure whose value is to be determined.

The ratio between two principal stresses can readily be determined by means of Mohr's Circle. In the dry granular soil considered, no resultant stress (on any plane) can have a greater obliquity than the angle of shearing resistance ϕ. *OR* and *OQ* (Fig. 7.1) therefore represent the limiting tangents within which the Mohr's Circle must be inscribed. Any circle fulfilling this condition gives the positions of P_1 and P_3. The ratio of the maximum principal stress to the minimum is

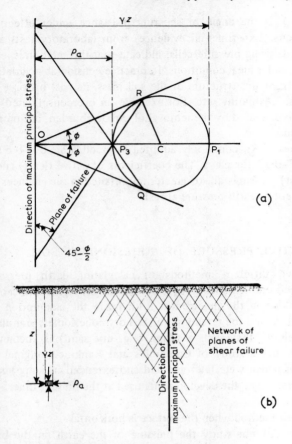

Fig. 7.1 Cohesionless soil: Rankine's solution for active pressure (horizontal surface)

then OP_1/OP_3; thus OP_3 represents the horizontal earth pressure to the same scale as that to which OP_1 represents the weight γz.

$$\sin \phi = \frac{RC}{QC} = \frac{\frac{1}{2}(\text{difference of principal stresses})}{\frac{1}{2}(\text{sum of principal stresses})}$$

$$\text{or} \quad \frac{p_a}{\gamma z} = \frac{1 - \sin \phi}{1 + \sin \phi} = \tan^2\left(45° - \frac{\phi}{2}\right).$$

Therefore

$$p_a = \gamma z \tan^2 \left(45° - \frac{\phi}{2}\right).$$

This may be written as

$$p_a = K_a \gamma z.$$

The total thrust per foot run on a vertical wall of height H is the mean pressure multiplied by the height; that is,

$$P_a = \tfrac{1}{2} K_a \gamma H^2.$$

OR represents the resultant with maximum obliquity, and it acts on the plane on which shear failure takes place. The direction of this plane is given by joining RP_3 (or QP_3), and, from the geometry of the figure, this plane makes an angle of $45° - \phi/2$ with the direction of the maximum principal stress. It can be imagined that if the soil is allowed to expand slightly, planes of failure such as are shown in Fig. 7.1b will develop.

Rankine's method for an inclined surface
Here again (Fig. 7.2) Rankine's method is to consider the equilibrium of an elemental portion of the material at depth z. The weight of the soil acts vertically, and the earth pressure considered is the stress which is conjugate to the vertical weight. The earth pressure on a vertical plane thus acts parallel to the surface, an assumption which leads to some inconsistencies and errors when applied to retaining walls.

This time the direction of the maximum principal stress is not known, and the Mohr's Circle diagram cannot be drawn in its correct orientation at the first attempt. For convenience the principal stress axis is drawn horizontally (fig. 7.2a) and OV and OH set off at β to the line OP_1, β being the obliquity of the resultant stresses on the conjugate planes considered (Fig. 7.2c). OV represents the known value of the resultant $\gamma z \cos \beta$, and a circle passing through V and tangent to the limiting lines OR and OQ locates the points P_1 and P_3. OH represents the value of the active pressure p_a and the angle VP_3H represents the angle between the planes on which $\gamma z \cos \beta$ and p_a act. When the Mohr's Circle diagram is in its correct orientation, VP_3

111

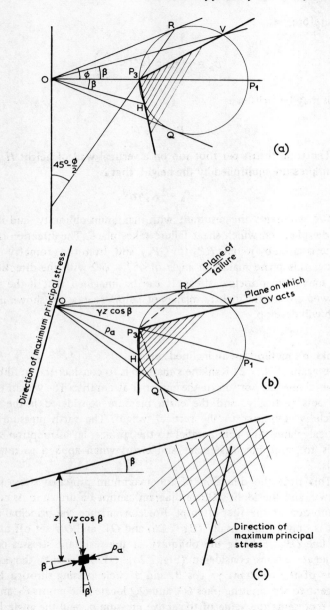

Fig. 7.2 Cohesionless soil: Rankine's solution for active pressure (inclined surface)

should be parallel to the surface of the soil, and P_3H should be vertical. Fig. 7.2b shows the diagram correctly orientated and indicates the directions of the planes of failure (P_3R and P_3Q). These planes still make an angle of $45° - \phi/2$ to the direction of maximum principal stress.

If this problem is solved analytically, as was done by Rankine, the ratio of P_a to $\gamma z \cos \beta$ is found to be

$$\frac{\cos \beta - \sqrt{(\cos^2 \beta - \cos^2 \phi)}}{\cos \beta + \sqrt{(\cos^2 \beta - \cos^2 \phi)}} .$$

The maximum value which is possible for β is ϕ. The stresses p_a and $\gamma z \cos \beta$ are then equal. It will be noted that in the above treatment only the pressure on a vertical plane is considered.

The wedge theory for cohesionless soil
This method, first developed by Coulomb, is of wider application than Rankine's method, and has been modified by various writers and experimenters to suit the conditions encountered in practice.

Instead of considering the equilibrium of an element within the mass of the material, Coulomb studied the whole of the material supported by a retaining wall when the wall is on the point of moving slightly away from the filling. If the wall is entirely removed, the material will finally settle at the angle of shearing resistance. The material considered is still a dry granular sand, and *BD* (Fig. 7.3a) represents its final plane of repose. When the wall moves forward only slightly, however, the surface of failure will not form immediately at the angle ϕ but will develop somewhere between the wall and the plane *BD*, at say, *BC*. For active pressure this surface of rupture *BC* has been shown by experiment to be very nearly a plane, and in the usual two-dimensional treatment *BC* is always drawn straight.

If this theory of the mechanism of failure is accepted, it is obvious that the pressure on the wall is caused, indirectly, by the weight of the wedge *ABC* which is held in equilibrium by the three forces shown in Fig. 7.3b. These forces are: the weight W of the wedge *ABC*, the resistance P_a offered by the wall, and the reaction R on the plane *BC*. Since the wedge tends to slip down behind the wall, the forces P_a and R act upwards in resistance to this movement and have their maximum possible obliquity. R acts at the angle of internal friction ϕ to the normal to *BC* and P_a at the angle of wall friction to the normal

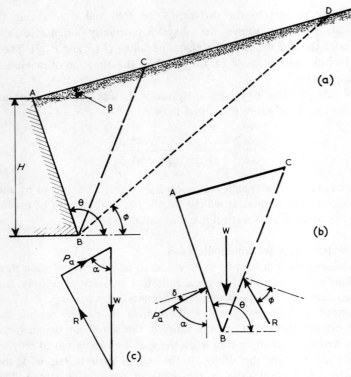

Fig. 7.3 Cohesionless soil: Coulomb's wedge for active pressure

to the back of the wall. The latter angle, δ, cannot exceed ϕ, but may be somewhat smaller, depending on the type of wall.

The problem, therefore, is to determine the position of the plane of rupture BC for which the wedge ABC will produce the maximum pressure on the wall.

For uniform soil this can be found analytically. Coulomb obtained the following expression for the active pressure p_a at depth z:

$$p_a = \gamma z \; \frac{\sin^2 (\theta - \phi)}{\sin^2 \theta \, \sin (\theta + \delta) \left[1 + \sqrt{\dfrac{\sin (\delta + \phi) \sin (\phi - \beta)}{\sin (\theta + \delta) \sin (\theta - \beta)}} \right]^2} \; .$$

This may be written as:

$$p_a = K_a \gamma z H^2 \sec \delta$$

where K_a is the coefficient of active pressure, i.e. the ratio of the normal component of pressure to γz.

The total thrust P_a on a vertical wall of height H is therefore

$$P_a = \int_0^H p_a dz = \tfrac{1}{2} K_a \gamma H^2 \sec \delta .$$

Graphical methods for the wedge theory
Several graphical constructions have been evolved for determining the maximum active thrust on a retaining wall. Of these, the most familiar are those known respectively as Rebhann's and Culmann's constructions, but they are limited to conditions of uniform soil properties and a plane upper surface, horizontal or sloping.

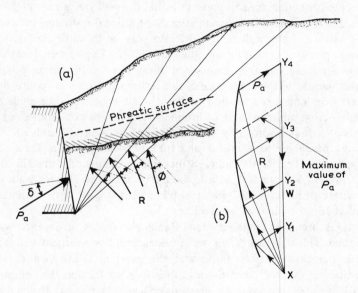

Fig. 7.4 Active pressure by trial wedges

The *method of trial wedges,* however, is simple in principle and is particularly useful where the conditions are not of the simplest and most uniform kind, for example, when the upper surface of the retained soil is irregular, or when the material is in layers having different bulk densities as in Fig. 7.4a.

Several planes of rupture are chosen, and the weights W of each of the wedges are calculated. These weights are set off upwards from the

115

point X (Fig. 7.4b) to points Y_1, Y_2, etc. The reaction R acts at an angle ϕ to the normal to the plane of rupture.

Vectors are drawn from X defining the directions of the reactions R on the various planes, and lines drawn from Y_1, Y_2, etc., parallel to the direction of P_a. By drawing a curve through the vertices of the force triangles the maximum value of P_a is determined for use in the design of the wall.

Wall friction

The question of whether wall friction should be taken into account when computing earth pressure has always been the subject of controversy. As, however, the basic assumption in the generally accepted theories of earth pressure is that the wall yields very slightly, it seems as reasonable to assume that the frictional resistance between the earth and the wall is brought into play as to allow for friction along planes of rupture in the soil itself. Experimental results generally bear out this assumption though it is difficult to obtain exact data as to the magnitude of this resistance. Packshaw points out there does not appear to be any direct relation between the angle of wall friction and the angle of internal friction. For cohesionless soils he states that δ probably has some value between 17 and $30°$. It seems reasonable to assume a value for δ between $\frac{1}{2}$ and $\frac{3}{4}\phi$. The Code of Practice 'Earth Retaining Structures' recommends that in the absence of definite test data δ should be taken as $20°$ for walls of concrete or brick, as $30°$ for steel piling coated with tar or bitumen, and as $15°$ for uncoated steel piling.

It is frequently stated that Rankine's theory disregards wall friction. This statement is apt to be misleading. For a vertical wall face with the earth surface horizontal the pressure is assumed to act normal to the wall, and there is therefore no friction. For sloping earth surfaces, however, the pressure on a vertical plane is assumed to act parallel to the earth surface, and therefore has both normal and shear components. When the vertical plane in the earth is replaced by a wall surface, the shearing resistance of the soil on this plane must be replaced by friction along the wall face. If the inclination of the earth surface, β, is less than the angle of wall friction, δ, this frictional resistance is of course less than the limiting friction.

If wall friction is taken into account, the magnitude of the resultant thrust is reduced. Also, its inclination is increased, further reducing the overturning moment on the wall.

For a vertical wall with the earth surface horizontal a simple appoximate method is to compute the thrust by Rankine's method and to multiply it by a correction factor to give the horizontal component of the thrust allowing for wall friction. The actual thrust is obtained by dividing this component by cos δ, and is assumed to act at an angle δ with the normal to the wall.

The correction factor varies only slightly with δ and the following values are a fair approximation.

Angle of wall friction	*Ratio K_a (wedge)* *K_a (Rankine)*
15°	0·90
20°	0·86
25°	0·83
30°	0·80

Referring to the formula for earth pressure according to the wedge theory, if $\theta = 90°$ (vertical wall) and $\beta = \delta$ (earth pressure acting parallel to the surface), the results obtained by Rankine's theory and those obtained by the wedge theory are identical.

Effect of level of water table

In the preceding sections the bulk density of the soil has been used in determining earth pressure. The effective weight of the soil varies, however, with the amount of moisture present, and with the position of the water table.

It is often assumed that above the water table the soil is completely saturated, and its bulk density therefore comprises the weight of water contained. Below the water table the water is under hydrostatic head and the submerged weight of the soil is its bulk density less the weight of unit volume of water. This decreases the earth pressure, but the water below the water table is under gravitational forces and acts through the voids independently of the soil. The pressure on the wall is thus increased by the normal hydrostatic pressure.

To summarize, above the water table the pressure is taken as being caused by the bulk density of the soil. Below the water table there is the pressure arising from the submerged density of the soil together with the independent hydrostatic pressure corresponding to the position of the water table.

Rain or spring water may, of course, be drained away rapidly because of the porosity of the soil or because of artificially constructed drains. If the water does not reach the back of the wall no water pressure is developed, but full hydrostatic head must be allowed for below the level of drains or weep holes.

7.2 ACTIVE PRESSURE OF COHESIVE SOILS

In cohesive soils the angle of shearing resistance under conditions of limited drainage is generally taken as zero, and the cohesive strength is therefore independent of the forces acting within the material. Further, since cohesive soils are capable of sustaining only a small tensile stress, cracks develop at the upper surface of the filling behind a retaining wall. The action of cohesive forces and the appearance of tension cracks are two factors which influence considerably the determination of the active pressure of cohesive soils. The two methods generally used for this purpose are modifications of those previously described for calculating the active pressure of cohesionless soils.

Bell's solution
The application to cohesive soils of the principle of conjugate stresses is shown in Fig. 7.5. The general case of a soil possessing both

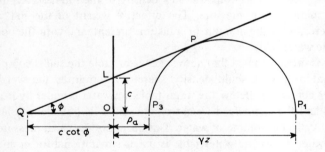

Fig. 7.5 Cohesive soil: Mohr's Circle solution for active pressure

cohesion and an angle of shearing resistance $\phi > 0$ is considered. Since the cohesion c is assumed to have a fixed value for a given soil, the Mohr's Circle diagram must be drawn to scale. The results obtained depend on scaled dimensions and not merely on ratios of dimensions as was the case with cohesionless soils. The first attempt to produce a

118

theory for the pressure of cohesive soil was made by Bell and published in 1914.

In Fig. 7.5 the limiting tangent to the Mohr's Circles is *LP*. This line produced back meets the axis at a distance $c \cot \phi$ from O. When the earth surface is horizontal OP_3 represents the active pressure p_a and OP_1 the vertical pressure γz.

The ratio

$$\frac{QP_3}{QP_1} = \frac{c \cot \phi + p_a}{c \cot \phi + \gamma z} = \frac{1 - \sin \phi}{1 + \sin \phi} = \tan^2 \left(45° - \frac{\phi}{2} \right).$$

From this, putting

$$\tan^2 (45° - \phi/2) = K_\phi$$

$$p_a = K_\phi \gamma z + (K_\phi - 1) c \cot \phi.$$

Put

$$K_\phi - 1 = \frac{1 - \sin \phi}{1 + \sin \phi} - 1 = - \frac{2 \sin \phi}{1 + \sin \phi}.$$

Thus

$$(K_\phi - 1) \cot \phi = \frac{-2 \sin \phi \cos \phi}{(1 + \sin \phi) \sin \phi} = - \frac{2 \cos \phi}{1 + \sin \phi}$$

$$= - 2 \tan \left(45° - \frac{\phi}{2} \right) = - 2 \sqrt{(K_\phi)}.$$

Therefore

$$p_a = K_\phi \gamma z - 2c \sqrt{(K_\phi)}.$$

Thus, comparing two soils, one cohesive and the other cohesionless, having the same bulk density and angle of shearing resistance, the cohesive soil, being more self-supporting, exerts a smaller pressure on the retaining wall than does the cohesionless soil, the difference being $2c \sqrt{(K_\phi)}$.

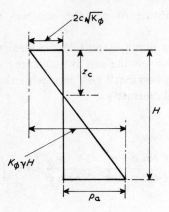

Fig. 7.6 Cohesive soil: pressure distribution on a vertical wall

The pressure distribution on a vertical wall is as shown in Fig. 7.6. Theoretically, down to a depth z_c the active pressure is negative, i.e. tensile. This critical depth, at which pressure begins to act on the back of the wall, is obtained by putting

$$K_\phi \gamma z_c = 2c\sqrt{(K_\phi)}$$

or

$$z_c = \frac{2c}{\gamma\sqrt{(K_\phi)}} = \frac{2c}{\gamma} \tan\left(45° + \frac{\phi}{2}\right).$$

When the angle of shearing resistance of the clay is zero the critical depth is $2c/\gamma$.

It should be assumed that there can be no tension between the wall and the soil down to a depth z_c. The active thrust per unit length of wall is then the area of the lower triangle in Fig. 7.6, that is,

$$P_a = \tfrac{1}{2}K_\phi \gamma (H - z_c)^2.$$

The thrust can also be obtained by integrating the expression for p_a between the limits $z = z_c$ and $z = H$. The thrust is then

$$P_a = \tfrac{1}{2}K_\phi \gamma (H^2 - z_c^2) - 2c^2\sqrt{(K_\phi)}(H - z_c)$$

which can be proved to be identical with the expression for P_a given above.

The wedge theory for cohesive soil

The solutions so far obtained for cohesive soils do not take into account the cohesion or friction on the back of the wall and are therefore of limited application. The wedge theory can be applied over a much wider field, and results obtained graphically to a sufficient standard of accuracy.

It is assumed that there is a neutral or ineffective zone of depth

$$z_c = \frac{2c}{\gamma} \tan \left(45° + \frac{\phi}{2} \right)$$

within which there is no adhesion or friction along the back of the wall or along the plane of rupture (Fig. 7.7).

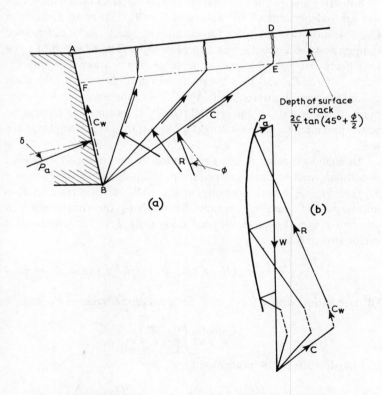

Fig. 7.7 Cohesive soil: active pressure by polygon of forces

121

There are five forces acting on the wedge of soil which is causing the pressure on the back of the wall:

  (i) the weight of the whole wedge *ABED (W)*;
  (ii) the reaction on the plane of rupture *(R)*;
  (iii) the resultant pressure on the wall *(P_a)*;
  (iv) the cohesion along the length *BE (C = c.BE)*;
  (v) the cohesion along the length of wall *BF (C_w = c_w. BF)*.,

The cohesion per unit area of wall c_w cannot be greater than *c*, and may be less, especially with hard clays. The Code of Practice suggests that where *c* is less than 50 kN m^{-2}, the value of c_w should be taken as equal to *c*, but where *c* exceeds 50 kN m^{-2}, c_w should be limited to this value.

Both (iv) and (v) are estimated from shear tests on the materials and are independent of the values of *R* or P_a. There are thus, as in Fig. 7.4, only two unknown forces, both of which can be determined in magnitude and direction by a force polygon. The selection of a few trial planes of failure allows of a curve being drawn defining the magnitudes of P_a, and thus the maximum possible value for the given conditions can be interpolated. Again, it is unnecessary to know the point of application of the resultant *R*. P_a is assumed, as is usual for active pressure, to act at a height of one-third of the height of the wall.

An analytical solution can be obtained for the following simplified conditions: wall vertical, soil surface horizontal, and $\phi = 0$. The force P_a (see Fig. 7.7) acts normally to the wall, and the reaction *R* is normal to the plane of rupture. By resolving the forces along an assumed plane of rupture inclined at an angle ψ to the horizontal it can be shown that

$$P_a = \gamma \frac{(H^2 - z_c^2)}{2} - c_w(H - z_c) \tan \psi - c(H - z_c) \operatorname{cosec} \psi \sec \psi.$$

Differentiation with respect to ψ for a maximum value of P_a leads to

$$\cot \psi = \sqrt{\left(1 + \frac{c_w}{c}\right)}$$

and substitution in the expression for P_a gives:

$$P_a = \gamma \frac{(H^2 - z_c^2)}{2} - 2c(H - z_c) \sqrt{\left(1 + \frac{c_w}{c}\right)}.$$

Drainage conditions in cohesive soil

During the backfilling of a retaining wall with cohesive soil the change in moisture content is likely to be very small. Thus for analysis of the stability during and shortly after the construction period the undrained shear strength is appropriate. Total stresses must be used in the calculations, that is the full weight of soil must be allowed for without deduction for pore-pressure. In this case, if the clay is fully saturated, there is the further simplification that $\phi = 0$.

In clay soil it is sometimes necessary to allow for vertical cracks which may be several feet in depth and which are likely to be filled with water. The head due to this water must be taken into account in estimating the gross pressure on the back of the wall.

For long-term stability, effective stress conditions should be assumed. If a drainage blanket is provided at the back of the wall a flow net should be drawn, from which the pore-pressures can be estimated. As the strength of soil normally increases with decrease in moisture content, the undrained analysis is usually on the side of safety.

7.3 PASSIVE RESISTANCE OF COHESIONLESS SOIL

In studying active pressure it was accepted that a slight movement of the wall in a direction away from the filling brought into play the maximum active thrust P_a. Similarly, a slight movement towards the filling evokes a resistance to such movement. This is known as the *passive resistance* (P_p), and it is always numerically greater than the active thrust for similar conditions of soil depth and density. Passive resistance occurs, for example, on the front face of a retaining wall which is tending to move slightly under the influence of active pressure on the back.

Rankine's method

As in Fig. 7.2 the problem is to find the value of the pressure acting parallel to the surface of the soil. This passive pressure (p_p) must be sufficient to overcome the downward thrust of the weight of the soil and to cause planes of shear failure to develop. Except when the surface is level, the direction of the maximum principal stress is not at first known, and for convenience is set out horizontally (Fig. 7.8a).

Lines OH and OV drawn at an angle β to the axis OP define the points on the circle representing the conjugate stresses. OV is the

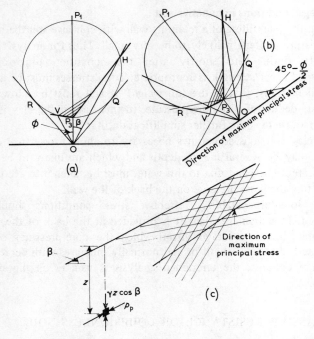

Fig. 7.8 Cohesionless soil: Mohr's Circle solution for passive resistance

known value of the vertical pressure $\gamma z \cos \beta$, and a circle passing through V and touching the lines of limiting shear, OR and OQ, defines the values of the principal stresses OP_1 and OP_3.

OH represents the value of the passive pressure intensity (p_p) and the direction P_3H represents the direction of the plane on which the passive resistance acts. This plane is known to be vertical, and when the whole Mohr's Circle diagram is rotated until P_3H is vertical (Fig. 7.8b) the correct orientation is obtained and the direction of the maximum principal stress defined. The planes of shear failure still make an angle of $45° - \phi/2$ with the direction of maximum principal stress, and the network of shear failures which develop on slight compression of the soil can be sketched (Fig. 7.8c).

Rankine's analytical solution for passive pressure when the upper surface is horizontal gives

$$p_p = \gamma z \tan^2 \left(45° + \frac{\phi}{2}\right)$$

or

$$p_p = K_p \gamma z.$$

If there is a sloping upper surface (at β to the horizontal), Rankine's solution gives

$$\frac{p_p}{\gamma z \cos \beta} = \frac{\cos \beta - \sqrt{(\cos^2 \beta - \cos^2 \phi)}}{\cos \beta - \sqrt{(\cos^2 \beta - \cos^2 \phi)}}.$$

Coulomb's expression for passive pressure when the upper surface is inclined is

$$\frac{p_p}{\gamma z} = \frac{\sin^2 (\theta - \phi)}{\sin^2 \theta \sin (\theta + \delta) \left[1 - \sqrt{\dfrac{\sin (\delta + \phi) \sin (\phi - \beta)}{\sin (\theta + \delta) \sin (\theta - \beta)}}\right]^2}.$$

The passive wedge

Coulomb's assumption that the surface of rupture behind a retaining wall is a plane can be extended to passive resistance calculations in cohesionless soil, provided (as has been pointed out by Terzaghi) that the angle of wall friction δ is not greater than $\phi/3$. When there is considerable friction between the wall and the filling, and δ increases beyond $\phi/3$, the sliding surface under passive resistance cannot be assumed to be even approximately plane.

If the surface is assumed to be plane (δ being small), the determination of the minimum passive resistance required to produce plastic failure in the filling may be determined by drawing a series of triangles of forces (as in Fig. 7.7) for various assumed positions of the plane of rupture (Fig. 7.9). It should be remembered that when the wall moves towards the filling the wedge tends to move upwards. The resisting reactions P_p and R must therefore appear on the upper sides of the normals (compare the effect of active pressure, Figs. 7.3 and 7.7).

The ϕ-circle method for cohesionless soil

When $\delta > \phi/2$ Coulomb's assumption that the slip surface under passive resistance is a plane is not valid. The lower portion of the slip surface is definitely curved.

For an ideal plastic material it can be shown that the shape of the slip surface is part of a logarithmic spiral but in practice it is generally considered as the arc of a circle (Fig. 7.10).

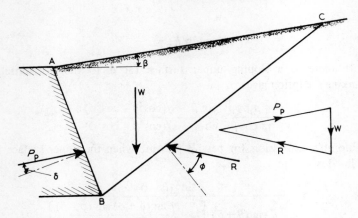

Fig. 7.9 Cohesionless soil: passive resistance by wedge theory

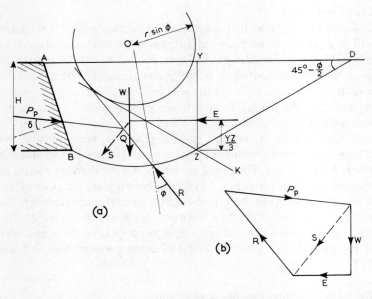

Fig. 7.10 Cohesionless soil: passive resistance by ϕ-circle

Whatever the shape of the lower portion, the upper part of the
wedge fails as indicated by Rankine, the planes of failure making
angles of $45° - (\phi/2)$ with the direction of maximum principal stress,

126

which is horizontal. If a line AZ is drawn at $45° - \phi/2$ to the horizontal upper surface, and a vertical YZ drawn from Z, then the effect of the portion YDZ may be represented by Rankine's passive pressure:

$$E = \tfrac{1}{2}\gamma(YZ)^2 K_p.$$

The problem thus resolves itself into a study of the equilibrium of the portion of the block $ABZY'$ In the *friction circle or ϕ-circle* method the curve BZ is assumed to be an arc of a circle.

The slip surface must be continuous, and ZD must therefore be tangent to the circular arc BZ. The centre of this arc lies on a line ZO (at right-angles to ZD) and in such a position that the circle passes through B.

At the point of failure the full frictional resistance of the cohesionless soil has been mobilized, and the resisting force at any point on BZD must act at ϕ to the normal. Along the curved portion the normals pass through the centre of the circle, and the resisting forces must thus all be tangent to a smaller circle whose radius is $r \sin \phi$. This small circle is called the *friction circle* or *ϕ-circle* (Fig. 7.10a)

It is now assumed that the resultant of all the resisting forces is also tangent to the ϕ-circle. This is not strictly true, but is sufficiently close to the truth to justify its acceptance.

The procedure is as follows (Fig. 7.19a):

(i) Measure the height of the vertical line YZ and calculate E, the earth pressure applied by the wedge YDZ, from $E = \tfrac{1}{2}K_p\gamma (YZ)^2$.

(ii) Measure the area $ABZY$ with a planimeter and calculate the weight W. The centroid of this area is also required to give the line of action of W. The simplest method of locating this is by cutting out a cardboard template and suspending it freely from several points in turn. Vertical lines through these points (as defined by a plumb-bob) all pass through the centroid which can thus be located. The use of an integrator is more exact, but such an instrument is not often available.

(iii) Combine these graphically to give the resultant S.

(iv) The forces acting on the portion $ABZY$ have now been reduced to three: S, which has just been found; R, the resultant reaction on the slip surface BZ, and P_p, the passive thrust on the back of the wall. These forces must all act

through one point Q determined graphically by the intersection of P_p (acting at δ to the normal to the wall) and S.

(v) Draw a line through Q tangent to the ϕ-circle. This is the
direction of the resultant R.

(vi) Draw P_p and R on the force diagram and scale off their
magnitudes.

This procedure is repeated for several other assumed slip surfaces.
For each of these the point Z, where the curved surface runs into the
straight, lies on the line AK (Fig. 7.10a). The least value of P_p which
can be determined is the value to be used in subsequent calculations.

7.4 PASSIVE RESISTANCE OF COHESIVE SOIL

When the wall is forced towards a cohesive filling a resistance is
developed as described for cohesionless soils. This passive resistance is
greater than the active pressure which accompanies a movement away
from the filling.

It has been shown that if the soil possesses cohesive strength the
active pressure is less than that produced by a cohesionless soil of a
comparable bulk density. Similarly, if a clay offers passive resistance
to compression, that resistance is greater than the passive force
developed in the equivalent cohesionless soil.

Mohr's Circle
Bell's solution for the value of the active pressure of cohesive soils was
developed by considering the Mohr's Circle of Fig. 7.5. In a similar
manner Bell's expression for the passive resistance of cohesive soil can
be found:

$$p_p = \gamma z \tan^2\left(45° + \frac{\phi}{2}\right) + 2c \tan\left(45° + \frac{\phi}{2}\right).$$

This applies only to the passive resistance of cohesive soils under
the simplest conditions, when the upper surface is horizontal and
there is assumed to be no friction on the back of the vertical wall. For
the more general cases other methods are employed, in which the
tendency of the soil to develop a slip surface is investigated.

The ϕ–circle method for cohesive soil
Again, a trial surface is assumed, consisting of a circlar arc BZ running

into a straight line *ZD* as in Fig. 7.10. The problem is worked out in three stages:

   (i) the determination of the resistances to displacement of the wedge *YZD*, namely, E_c due to cohesion and E_w due to friction;

  (ii) the determination of the resistance P_c of the block *ABZY* under the action of E_c and the cohesion forces along the slip surface and along the back of the wall;

 (iii) the determination of the resistance P_w caused by the weight of the block *ABZY* and the force E_w.

The total passive resistance $P_p = P_c + P_w$.

The calculations must be repeated for several trial slip surfaces in order to find the minimum value of P_p.

A worked example of this procedure is given in *Problems in Engineering Soils* by Capper, Cassie and Geddes.

7.5 WALLS WHICH DO NOT FAIL BY ROTATION ABOUT THE LOWER EDGE

In all the investigations in the earlier part of this chapter it was assumed that the walls failed by a slight rotation about a point in the lower edge (*B* in the diagrams). Certain types of earth-retaining structures are supported in such a manner that they cannot fail by simple rotation. Into these categories come the timbering of trenches, cantilever sheet-pile walls, and anchored bulkheads, each of which tends to support the earth pressure in a manner not susceptible to treatment by the methods already discussed. A brief treatment of each of these types is given in this article.

Slip surfaces in timbered trenches

If the timbering of a trench fails, allowing the supported material to slip, the slip surface is quite different from that behind a wall which fails by rotation about its lower edge. Behind the timbering at the top the filling cannot move outwards because of the customary strutting. The wedge must therefore move vertically downwards at this level, and the slip surface cuts the upper soil surface at $90°$. Again, the lower ends of the timbers are not well supported laterally, and any relatively large movement which takes place occurs at the bottom of

129

Pressures applied by a mass of soil

the trench. It can thus be estimated that points on the slip surface will be nearer the face of the support at the top of the trench and farther away at the bottom than in the Coulomb wedge. Fig. 7.11 shows the type of curve which can be expected and compares it with the Coulomb wedge. Terzaghi found that this slip surface can be represented by a logarithmic spiral.

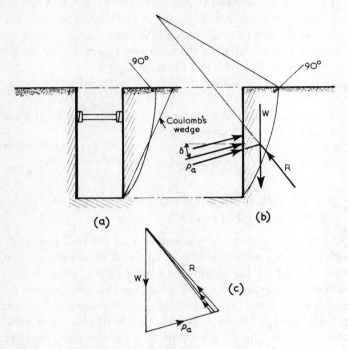

Fig. 7.11 Earth pressure on trench timbering

The methods employed to find the magnitude of the maximum active pressure are similar to those of Figs. 7.4 and 7.7, if it is remembered that the direction of the resultant R is defined by a line joining the intersection of P and W with the centre of the logarithmic spiral.

When a flexible wall bends under the action of the active earth pressure the full frictional resistance of the filling is mobilized towards the centre of the height of the wall because of the deflection of the wall at that point. The filling there may thus slip long before there is any sign of failure of the wall itself. Such slipping of the central part of the filling throws the active pressure load towards the top and the bottom of the wall, which have not moved. Because of this tendency to throw load towards the 'abutments', this characteristic is known as *arching*. Its effect is to change the distribution of pressure from the hydrostatic (with the resultant at one-third of the height) to one which is more symmetrical about the centre of the depth. The result of this is that the active pressure P_a acts at some point near the middle of the depth. In sands and silts Terzaghi found the resultant active pressure to act at 0·45 to 0·55h from the bottom of the trench.

In cohesive soils the height at which the resultant active pressure acts show more variation and depends on the height at which the side of the trench would stand temporarily without timbering. The other variables – c, γ, ϕ, etc. – also have an effect, but in general it may be assumed that P_a, even in cohesive soils, acts at about half the height of the supported face. The diagram shows three trial positions of the force P_a. Fig. 7.11c shows that the magnitude of the active thrust is not much affected by this variation.

Cantilever sheet piling
When the height of earth to be retained by sheet piling is small the piling acts as a cantilever, as shown in Fig. 7.12a. The whole of the active earth pressure acting for the height $h + d$ on the right must then be resisted by the development of passive resistance in the soil on both sides of the wall. The moment of these passive resistances constitutes the necessary fixing moment for the cantilever. For both active and passive pressures it is usual to assume linear variation with depth. If the wall is flexible this assumption is not correct, but the required depth of driving, which is of primary importance in cantilever sheet piling, is not seriously affected by changes in pressure distribution.

Figs. 7.12b and c show the assumed distribution of active and passive pressures on the wall. For simplicity it is assumed that the soil on both sides of the wall is of uniform density γ and that there is no hydrostatic pressure. Variations in density and the presence of a water table can be allowed for. Suppose the wall to undergo very slight angular rotation about some point C at a distance x above the foot of the wall, D. Passive resistance will be mobilized in front of the wall down to the level of C, and also behind the wall below this point. The minimum driving depth d for equilibrium can be determined by

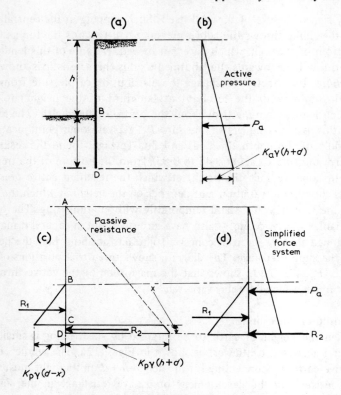

Fig. 7.12 Earth pressure on cantilever sheet piling

solving simultaneously two equations in which x and d are the unknowns. One equation is formed by equating to zero the algebraic sum of the active and passive thrusts, and the other by taking moments of these thrusts about the foot of the wall. The solution of these equations is very laborious, and the problem can be considerably simplified by assuming the passive reaction R_2 as a concentrated force acting at the foot of the pile. The simplified arrangement is shown in Fig. 7.12d. For equilibrium the moments of the active pressure on the right and the passive resistance on the left about the point of action of R_2 must balance.

$$\frac{R_1 d}{3} - \frac{P_a(h+d)}{3} = 0.$$

Now

$$R_1 = \tfrac{1}{2}K_p\gamma d^2$$

and

$$P_a = \tfrac{1}{2}K_a\gamma(h + d)^2$$

Therefore

$$K_p d^3 - K_a(h + d)^3 = 0.$$

The solution of this equation gives a value for d which is at least a guide to the driving depth required. The depth calculated should be increased by at least 20 per cent to provide a factor of safety and to allow extra length for the development of the passive force R_2.

The following approximate figures, applicable to cohesionless soil with no hydrostatic pressure, are given by Henry.

ϕ	d
$20°$	$2 \cdot 0h$
$25°$	$1 \cdot 5h$
$30°$	$1 \cdot 2h$
$35°$	$0 \cdot 9h$
$40°$	$0 \cdot 7h$

Sheet-pile walls of this kind should be used only in good frictional materials. If designed as described they will be of adequate dimensions. No account has been taken of the flexure of the wall, nor of the friction between wall and filling, which greatly increase the passive resistance available from frictional materials. Both these omissions tend to simplify calculation and may be regarded as bringing into play an added factor of safety.

Anchored bulkheads

When a sheet-pile wall is held near the top by a tie-bar anchored back into the stable volume of soil behind the plane of rupture, stability does not depend so much on the development of large passive pressures as it does in sheet-piling walls. The result is that the driving depth can be smaller in proportion to the depth of retained soil than it is in the cantilever sheet-pile wall. In addition, the total depth of retained soil can be greater for a given strength of sheet piling.

The principal methods of analysing the equilibrium of anchored piling are based on one of two assumptions as to the method of support of the driven end. These are known respectively as the *free earth support* and the *fixed earth support* assumptions. In the ensuing analyses the density of the soil is, as before, assumed to be uniform.

Free earth support. Here the wall is assumed to be freely supported, near the top by the anchorage force and below ground level by the passive resistance of the soil.

Let d be the minimum depth for stability, that is, with a factor of safety of unity. The loading is as shown in Fig. 7.13b.

$$R_1 = \tfrac{1}{2}K_p\gamma d^2$$

$$P_a = \tfrac{1}{2}K_a\gamma(h+d)^2.$$

Taking moments about A,

$$\tfrac{1}{2}K_p\gamma d^2(h_1 + \tfrac{2}{3}d) = \tfrac{1}{2}K_a\gamma(h+d)^2[\tfrac{2}{3}(h+d)-(h-h_1)]$$

$$= \tfrac{1}{2}K_a\gamma(h+d)^2(\tfrac{2}{3}d - \tfrac{1}{3}h + h_1).$$

This cubic equation in d is best solved by trial. In order to provide a factor of safety it may be assumed that only a fraction, say one-half, of the full passive resistance is to be mobilized. The value of K_p in the above equation may therefore be taken as half its ordinary value. Alternatively, the depth calculated on the assumption of full passive resistance may be increased by about 50 per cent.

The final value of d having been decided, the force in the anchor ties can be found by taking moments about the line of action of the passive reaction R_1.

$$R_A(h_1 + \tfrac{2}{3}d) = \tfrac{1}{2}K_a\gamma(h+d)^2[(h+d)-\tfrac{1}{3}d]$$

$$= \tfrac{1}{6}K_a\gamma(h+d)^2h.$$

The value of R_A thus found is, of course, the reaction per foot run of wall and the actual forces in the tie-rods will depend on their horizontal spacing.

Fixed earth support – graphical integration method. In this method the lower part of the wall is assumed fixed in a manner similar to cantilever sheet piling. First a driving depth d is assumed and the loading diagram is drawn, as shown in Fig. 7.14b. From this the bending moment diagram is constructed. Assuming no movement of

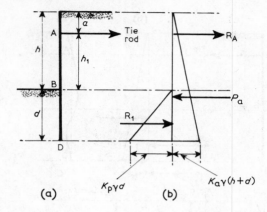

Fig. 7.13 Earth pressure on anchored bulkhead – free-earth support

the anchor tie the deflected shape of the wall must pass through the point A (Fig. 7.14c). By graphical integration of the bending moment diagram a trial elastic line is drawn, considering the wall direction-fixed at the foot. If the assumed depth is not correct this elastic line will not pass through A. Adjustment of the driving depth until this fault is rectified gives the correct value for d.

Fixed-earth support – equivalent beam method. The above procedure is tedious and a shorter, if less exact, solution can be formed by the equivalent beam method illustrated in Figs. 7.14d, e and f. The anchored wall has two points of contraflexure, and if the lower of these points, C is considered as a hinge, transmitting shear only, the structure may be regarded as two separate beams, AC and CD, the force R_c being the reaction at the hinge.
The procedure is as follows:

(a) The position of the point of contraflexure C is first chosen. The depth y of this point below ground surface may be found by interpolation from the following figures given by Terzaghi, which were determined by the graphical integration method:

ϕ	y
20°	$0 \cdot 25h$
30°	$0 \cdot 08h$
40°	$-0 \cdot 006h$

135

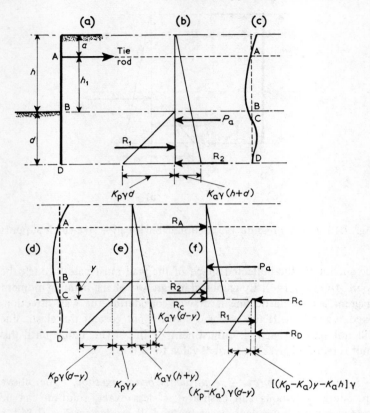

Fig. 7.14 Earth pressure on anchored bulkhead – fixed-earth support

The convenient approximation of $y = 0 \cdot 1h$ is applicable to many normal frictional soils.

(b) The forces on the upper beam are calculated from the active and passive pressures down to the level of C. These are:

$$P_a = \tfrac{1}{2} K_a \gamma (h + y)^2$$

and

$$R_2 = \tfrac{1}{2} K_p \gamma y^2.$$

(c) The reaction R_c at the 'hinge' C is found by taking moments of these forces about A:

$$R_c(h + y - a) =$$
$$\tfrac{1}{2} K_a \gamma (h + y)^2 \left[\tfrac{2}{3}(h + y) - a \right] - \tfrac{1}{2} K_p \gamma y^2 (h - a + \tfrac{2}{3} y).$$

(d) The loads on the beam *CD* are represented by trapezoidal pressure areas which can be divided into rectangles and triangles as shown in Figs. 7.14e and f. Taking moments about *D*

$$R_c(d-y) =$$

$$\tfrac{1}{2}(K_p - K_a)\gamma \frac{(d-y)^3}{3} + [(K_a - K_p)y - K_a h]\,\gamma\,\frac{(d-y)^2}{2}.$$

Hence

$$\frac{6R_c}{\gamma(K_p - K_a)} = (d-y)^2 + \left(3y - \frac{3K_a}{K_p - K_a}\,h\right)(d-y).$$

Solving this equation as a quadratic in $(d - y)$ we get

$$d - y =$$

$$\frac{3}{2}\left(\frac{K_a}{K_p - K_a}\right)h - y + \sqrt{\left[\frac{9}{4}\left(y - \frac{K_a}{K_p - K_a}\,h\right)^2 + \frac{6R_c}{\gamma(K_p - K_a)}\right]}$$

or

$$d =$$

$$\frac{3}{2}\left(\frac{K_a}{K_p - K_a}\right)h - \frac{y}{2} + \sqrt{\left[\frac{9}{4}\left(y - \frac{K_a}{K_p - K_a}\,h\right)^2 + \frac{6R_c}{\gamma(K_p - K_a)}\right]}.$$

The first term under the square root is very small compared with the second and may be neglected, giving

$$d = \frac{3}{2}\left(\frac{K_a}{K_p - K_a}\right)h - \frac{y}{2} + \sqrt{\left(\frac{6R_c}{\gamma(K_p - K_a)}\right)}.$$

The substitution in this expression of the assumed value of *y* and the value of R_c as found above enables the minimum driving depth *d* to be calculated.

As in the case of cantilever sheet piling the calculated driving depth should be increased by 20 per cent to allow for the fact that the lower passive resisting force R_D is not a knife-edge reaction as assumed in the simplified calculations. The *fixed-earth support* method of design results in a value of *d* greater than that found by the *free-earth support* method, hence the suggested addition of 20 per cent will usually provide an adequate factor of safety.

The Code of Practice 'Earth Retaining Structures' recommends the use of the *fixed-earth* hypothesis, since this has the effect of reducing the maximum bending moment in the sheeting, thus permitting a lighter section of piling to be used. On the other hand, extensive experiments at Manchester by Rowe support the validity of the *free-earth* method. Rowe suggests that when using this method wall friction should be allowed for in computing the active pressure, taking $\delta = \frac{2}{3}\phi$; for the passive resistance, wall friction should be neglected and a coefficient equal to two-thirds of the maximum K_p should be used.

Rowe has also investigated, both analytically and experimentally, the effect of the flexibility of the sheeting. He finds that the maximum bending moment may be reduced by as much as 25 per cent, depending on the flexibility which can be expressed in the form of a *stability number* defined as H^4/EI where:

H = total length of piling (that is, $h + d$);
E = modulus of elasticity of piling;
I = second moment of area of cross-section per unit length of wall.

7.6 APPLICATION OF THE WORK OF CHAPTER 7

It is often necessary to practice to assess at least the order of magnitude of the pressure on retaining wall and sheet piling and the methods given in this chapter, both by calculation and by graphical procedures, should be practised.

Application 7A
It is required to find the thrust on a wall 10 m high, supporting cohesionless material. The water table is 2 m below the surface. From the surface to 5 m below, the bulk density is $1.7\,Mg\,m^{-3}$. The material below this level has a bulk density of $1.85\,Mg\,m^{-3}$. K_a for both soils is 0.33.

(i) In determining the active pressure by Rankine's method, or in calculating the weights for the triangle of forces in the wedge method, the unit weights to be used are:

Upper material: Above the water table
$$1.7 \times 9.81 = 16.7 \text{ kN m}^{-3}$$
Below the water table
$$(1.7 - 1)9.81 = 6.9 \text{ kN m}^{-3}$$

Lower material: $(1.85 - 1)9.81 = 8.3$ kN m^{-3}.

(ii) The resulting active pressures are:

At 2 m depth $\quad 0.33 \times 16.7 \times 2 = 11.0$ kN m^{-2}
At 5 m depth $\quad 0.33 \times 6.9 \times 3 + 11.0 = 17.8$ kN m^{-2}
At 10 m depth $0.33 \times 8.3 \times 5 + 17.8 = 31.5$ kN m^{-2}.

(iii) Draw the pressures on the back of the vertical wall as horizontal lengths to scale, at 2, 5 and 10 m. The area of the diagram gives the total pressure on the wall in kN. This refers to the pressure of the soil only (the effective pressure).

(iv) The thrust P_w due to the water pressure acting on the part of the wall below the water table must then be calculated. The combined pressure is then $P_a + P_w$.

Application 7B
A vertical retaining wall 7.5 m high retains soil of a mass of 1.8 Mg m^{-3}. The shape of the surface of the soil is shown in Fig. 7.15, and the figure also shows two special rail tracks each carrying 100 kN m^{-1} of length. What percentage increase in maximum pressure on the wall is produced by the addition of the rail tracks? ϕ for the soil is 35° and the angle of friction on the back of the rough wall is 25°.

(i) This is clearly a problem for the use of the wedge method. Join A to C as an obvious first wedge, and then set out distances along the horizontal surface until the wedge produced shows a boundary AF which is too flat to be critical. After some experience, you will be able to estimate where the critical slip surface occurs. In this problem it probably lies somewhere close to AD, merely by visual inspection.

(ii) Calculate the weights of the wedges: (assume 1 m length of soil at right angles to the plane of the diagram).

Wedge ABC weighs $0.5 \times 7.5 \times 3 \times 1.8 \times 9.81$ kN = 200 kN
Wedges ACD, ADE, AEF are all $0.5 \times 9 \times 3 \times 1.8 \times 9.81$ kN = 240 kN.

To the weights of the wedges must be added the weights of the rail loading: 100 kN on the second wedge, and 200 on each of the last two.

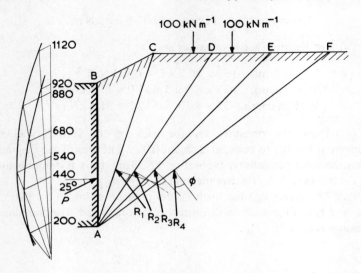

Fig. 7.15

Total weights to be used in analysis:

Wedge *ABC*   200 kN
Wedge *ABCD*   200 + 240 = 440 kN and + 100 line load =   540 kN
Wedge *ABCE*   440 + 240 = 680 kN and + 200 line load =   880 kN
Wedge *ABCF*   680 + 240 = 920 kN and + 200 line load = 1120 kN.

(iii) Only the pressure on the wall is required, so a triangle of
forces can be drawn to a suitable scale, showing the *weight* of the
wedge, which is known, the *direction* of *R* (at angle of shearing
resistance to the normal to the slip plane) and the *direction* of *P* (at
25° to the normal to the wall).

(iv) For each of the complete wedges, from the wall to the slip
plane, the triangle of forces shows the value of the active force *P* on
the wall. Joining these by a curve shows at what value force *P* can be
accepted as a maximum.

(v) The inner curve shows the variation of *P* without the surcharge
of the rail tracks, and the outer curve shows the variation of *P*, with
the surcharge in action.

(vi) A large scale drawing is required to make an accurate
assessment, but when that is done, the increase in pressure on the wall

due to the rail loading is in the region of 25 per cent. A vertical tangent to each of the curves locates the point to which the value of P should be measured (at an angle to the horizontal, in this case, of $25°$).

Application 7C

A horizontal rock surface is covered to a depth of 3100 mm with a layer of soft clay ($\phi_c = 4°$, $\gamma_c = 1·4\,Mg\,m^{-3}$, $c = 28·7\,kN\,m^{-2}$). A high embankment of granular material ($\phi_g = 30°$, $\gamma_g = 2·4\,Mg\,m^{-3}$) is to be constructed. To avoid risk of subsidence, the site is excavated to rock level before placing the granular material. Plot a curve showing how the intensity of pressure between the clay and the granular material at point X, varies with the angle (β) of the slope of the embankment. Secondly, determine what angle of slope would give a factor of safety of 4 against disturbance or slip of the soft clay (Fig. 7.16).

(i) The granular material applies an active force on the clay, and the clay produces a passive resistance. The criterion is that the passive resistance should not develop to more than one quarter of the active pressure of the granular fill.

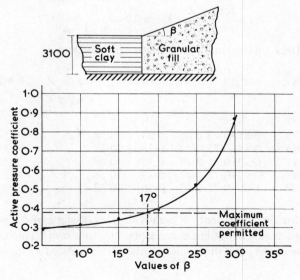

Fig. 7.16

141

(ii) Since the materials are homogeneous, and the state of loading is not complex, this is a situation in which the various mathematical expressions, describing the pressures in a mass of soil, can be used.

(iii) Use Rankine's expression for active pressure for cohesionless soil and Bell's expression for passive resistance of cohesive soil.

Active pressure on the vertical interface: (at X):

$$p_a = \gamma_g z \cos \beta \frac{\cos \beta - \sqrt{(\cos^2\beta - \cos^2\phi_g)}}{\cos \beta + \sqrt{(\cos^2\beta - \cos^2\phi_g)}}.$$

Passive resistance on the vertical interface (at X):

$$p_p = \gamma_c z \tan^2 \left(45 + \frac{\phi_c}{2}\right) + 2c \tan \left(45° + \frac{\phi_c}{2}\right).$$

The passive resistance can be calculated in numerical terms:

$$p_p = (1·4 \times 9·81)3·1 \tan^2 47° + 2 \times 28·7 \tan 47°$$

$$= 110 \text{ kN m}^{-2}.$$

The active pressure must not be greater than one quarter of this:

$$p_a = (2·4 \times 9·81)\,3·1 \cos \beta \, \frac{\cos \beta - \sqrt{(\cos^2\beta - \cos^2 30°)}}{\cos \beta + \sqrt{(\cos^2\beta - \cos^2 30°)}} \not> \frac{110}{4}$$

$$\cos \beta \, \frac{\cos \beta - Z}{\cos \beta + Z} = Y \cos \beta \not> \frac{110}{4 \times 73} \not> 0·377.$$

(iv) Draw up a calculation table for various values of

β	$\cos \beta$	$\cos^2\beta - \cos^2 30°$ $= Z^2/c$	Z	$Y = \dfrac{\cos \beta - Z}{\cos \beta + Z}$	$V = Y \cos \beta$
$30°$	0·87	zero	zero	1·0	0·87
$25°$	0·90	0·81−0·75 0·06	0·24	0·58	0·52
$20°$	0·94	0·88−0·75 0·13	0·36	0·45	0·39
$15°$	0·97	0·93−0·75 0·18	0·42	0·40	0·35
$10°$	0·99	0·97−0·75 0·22	0·47	0·36	0·31
$0°$	1·00	1·00−0·75 0·25	0·50	0·33	0·29

(v) Plot the coefficients V against degrees of slope of the granular bank, and read off the angle corresponding to a coefficient of 0·38. The slope of the bank should not be steeper than about $17°$.

CHAPTER 8

Stability of slopes

From the engineering point of view, the problem of the stability of slopes arises in the following conditions:

(i) the study of the mechanism of large-scale natural slips on hillsides and on cliff or mountain faces;

(ii) the construction of artificial slopes in cuttings and embankments for roads and railways;

(iii) the construction of earth dams and water-retaining embankments;

(iv) the application of remedial measures where slips have occurred, in order to repair damage and prevent extension.

Any slope is subject to the action of weathering agencies such as snow, frost, wind, and rain, and sometimes to the undercutting action of rivers or the sea. The combined effect of these agencies is gradually to flatten and lengthen the slope. If the scale is sufficiently large, such flattening movements, although gradual in the geological sense, may be catastrophically large and violent from the point of view of the engineer. Rock falls producing movement of scree slopes may destroy a roadway, the internal erosive action of water in a sandy stratum may cause subsidence and slips, or peat layers lubricated by melting snow may slide on the underlying rock.

The action of water is particularly productive of slope movements. Clay and shale softened by rain may ultimately form a shallow, weak layer on the surface of harder material. The shear strength of this material decreases as the moisture content increases, and the whole

143

layer is liable to slip as a detritus slide or mud run. Again, water percolating into fissured clay leads to progressive deterioration and weakening, and, eventually, the shear strength of the clay is so reduced that a rotational slip takes place.

Whether slips occur in large-scale natural slopes or in the smaller man-made cuttings and embankments, the same principles can be studied in them all. Slips sometimes occur quite suddenly; in other instances failure is gradual, commencing with cracks at the top of the bank and slight upheaval near the toe, later developing into a complete slip.

8.1 PRINCIPLES OF STABILITY

The disturbing force is the weight of the soil and the tendency to slip is restrained by the shearing resistance along the surface of rupture. A slip can thus be caused either by an increase in the disturbing force or by a reduction in the shearing resistance of the soil, or by a combination of both. Additional weight may be imposed by the erection of buildings, by the tipping of extra material on top of the bank, or by an increase of water content after heavy rain. Weakening of the soil may be due to an increase in pore pressure or, in some clays, to a progressive weakening with time. The profile of a slip is often governed by the presence of a harder stratum, which may be level or sloping, or by zones of weaker material.

Rotational slips
A common type of slip, and one which is amenable to quantitative investigation, is the rotational shear slip – a common occurrence in cohesive soil. Rotational slips are not surface phenomena like detritus or flow slides, but are deep-seated. A large mass of material forming the slope slips on a curve which commences beyond the top of the slope and finishes at or near the toe. The slip surface takes a form which, at least for a short length of the bank, is approximately cylindrical. If the total mass of the slipped material is considered, the slip surface is reminiscent of a conchoidal fracture.

Effective stress
Analysis of the stability of slopes may be carried out in two ways, using either

    (i) total stress and undrained shear stress parameters c_u and ϕ_u, or

(ii) effective stress and shear parameters c' and ϕ'.

The former may be used when considering short-term stability when there is little drainage and therefore little dissipation of pore-pressure. An example is the immediate stability of an embankment constructed of cohesive soil. In saturated clay, where $\phi_u = 0$, the total stress hypothesis leads to a simplified procedure known as the '$\phi = 0$ analysis'.

The effective stress analysis is used for long-term stability problems, and is particularly applicable to the condition known as rapid drawdown in earth dams. For this type of analysis it is necessary to know or to estimate the pore pressure in the soil. The effective stress method is now often used as an alternative to the total-stress method even where the latter method would be justifiable.

Effect of pore-pressure
The stability of an embankment is greatly influenced by the pore-pressure in the soil, which reduces the effective pressures on the slip surface, thus lowering the mobilized frictional resistance to movement. Problems concerning pore-pressure may be divided into two categories:

(i) where the pore-pressure is an independent variable, controlled by static ground-water level or by conditions of steady seepage;

(ii) where changes in pore-pressure are induced by changes in the stress in the soil.

An example of (i) is a mature bank in which the pore-pressure depends on the position of the phreatic surface or, during seepage, on the location of the equipotential lines of the flow net. To find the pore-pressure at the point A in Fig. 8.1, sketch an equipotential line through A to cut the phreatic surface at B. This indicates the level to which water would rise in a standpipe at A. The pore-pressure at A is, therefore, $\gamma_w(h + h_1)$. The *excess pore-pressure*, that is the difference between the recorded pore-pressure and the static pressure, is $\gamma_w h_1$. If seepage were taking place in a direction away from the impounded water, the phreatic line would be below the water in the reservoir and h_1 would be negative.

In the second category the pore-pressure changes with time, the

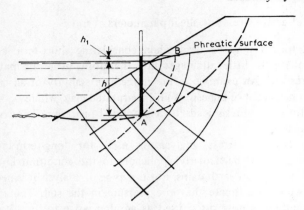

Fig. 8.1 Excess pore-pressure.

rate of adjustment to equilibrium depending on the permeability, the excess pore-pressure gradients, and the distance to a free-drainage surface.

Bishop has suggested the use of a dimensionless parameter, the pore-pressure ratio $r_u = u/\gamma z$, where z is the depth below the soil surface. It is often justifiable to assume homogeneous pore-pressure distribution, in which case r_u is taken as constant throughout the slope.

Factor of safety

In the effective-stress analysis it is assumed that failure along a slip surface is governed by the shear strength, $s = c' + (\sigma - u) \tan \phi'$, where σ is the normal stress on the surface of shearing. The factor of safety is defined as the factor by which the ultimate shear strength can be reduced before the slope is brought into a state of incipient failure.

Under conditions of equilibrium the mobilized shearing resistance is made up of cohesion and friction. Thus the shearing stress on a potential slip surface can be expressed as:

$$s = \frac{c'}{F_c} + (\sigma - u) \frac{\tan \phi'}{F_\phi}$$

where the letter F refers to factors of safety. For the sake of simplicity, it is assumed that the factor of safety is the same for both cohesion and friction.

On any slip surface there is always a variation in σ and u, and often also of c' and ϕ', from point to point. An alternative concept of the factor of safety is therefore often used, that is, to consider the moments of the forces acting about the centre of rotation. The disturbing moment is that due to the weight of the soil tending to slip; the restraining moment is that of the shearing resistance acting along the slip surface. The factor of safety is defined as the ratio of the maximum restraining moment to the disturbing moment.

The same reasoning applies when using the total stress analysis. Then c_u and ϕ_u are substituted for c' and ϕ', and u is neglected.

Residual strength

The question of the residual strength of certain types of clay is discussed in an earlier chapter. It seems probable that this phenomenon is the cause of many failures both in natural and man-made slopes after they have remained stable for many years. If the shear strength of the clay gradually decreases towards its ultimate or residual value, a time comes when the strength is insufficient to resist the disturbing forces, and failure results.

8.2 ANALYSIS OF STABILITY

Various methods of analysis have been suggested for investigating the stability of earth embankments and cuttings. These methods are similar in principle to the wedge theories of earth pressure. A surface of rupture is assumed and an investigation is made of the equilibrium of the mass of earth which tends to slip. By repeating the process for a number of possible surfaces of rupture, the most dangerous slip surface is found, that is the surface for which the factor of safety has its minimum value. The shape of the surface of rupture may be influenced by some geological feature. The assumption of a logarithmic spiral for the profile of the slip surface leads to an analytical solution, but the calculations are long and tedious. A circular arc, or linked circular arcs, are, therefore, generally used. The analysis of a sufficient number of trial slip surfaces to ensure locating the most dangerous is laborious, especially when the more elaborate methods of analysis are employed. It is now possible to manipulate the equations of equilibrium into a form suitable for solution by electronic computer.

Method of slices
In this method, originally devised by Fellenius, a slice of unit thickness, of the volume tending to slide is divided into vertical slices,

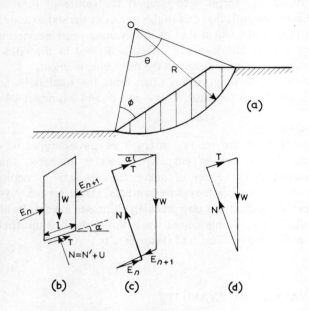

Fig. 8.2 Method of slices.

as shown in Fig. 8.2a. Any one of these slices of height h and breadth b, is in equilibrium under the action of the five forces shown in Fig. 8.2b. These forces are:

 (i) the weight of the slice, $W = \gamma h l \cos \alpha$;
 (ii) the normal reaction N on the slip surface, consisting of the inter-granular reaction N' plus the force U due to pore-pressure. If the pore-pressure ratio is r_u, then $U = r_u \gamma h l$;
 (iii) the tangential force T due to cohesive and frictional resistance mobilized on the slip surface

$$= \frac{c'l + N' \tan \phi'}{F}$$

 (iv) and (v) the inter-slice reactions E_n and E_{n+1}.

Of these forces, W is known in magnitude and direction, the directions of N and T are known, but the E forces are completely unknown in magnitude, direction, and point of application. The force diagram, Fig. 8.2c, is therefore, interdeterminate.

To obtain a complete solution, the following conditions must be satisfied:

(i) the forces acting on each slice must be in equilibrium;
(ii) the vectorial sum of the resultants of each pair of inter-slice reactions must be zero;
(iii) the sum of the moments of these resultants about any point in their plane must be zero.

By a process of successive approximations, a value of F is found which satisfies all the conditions. This is a long laborious procedure and, as it has to be repeated for a number of trial slip surfaces to evaluate the minimum F, the use of a computer is essential.

The Bishop method of slices

A less-demanding solution has been suggested by Bishop. The inter-slice forces are assumed to be horizontal. In other words, the vertical component of these forces on each slice are assumed to balance out. Trial and error are still necessary to find F, but only one equilibrium equation – that of moments – has to be solved. It is claimed that the error introduced by this simplifying assumption is unlikely to exceed one per cent.

The weight of each slice is obtained either by calculation or from planimeter measurement of the area. The inclination α of its surface to the horizontal is measured. Then, for any slice, by resolving the forces vertically,

$$W = N \cos \alpha + T \sin \alpha.$$

But

$$T = sl/F.$$

The normal pressure on the face of the slice is

$$\sigma = \frac{N}{l} = \frac{W}{b} - \frac{s \tan \alpha}{F}$$

149

thus

$$s = \frac{c + W/b \tan \phi}{1 + (\tan \alpha \tan \phi)/F} . \tag{1}$$

The E-forces are now horizontal, and the sum of their moments about O is zero, according to (iii) above. Taking moments about the centre of rotation O results in the equation

$$\Sigma Wx = R \Sigma T = R \Sigma(sl/F).$$

where x is the horizontal distance from W to centre O.

Thus
$$F = \frac{R \Sigma sl}{\Sigma Wx} .$$

Putting $l = b \sec \alpha$ and combining this with the expression for s (Equation 1 above) gives

$$F = \frac{1}{\Sigma W \sin \alpha} \sum \frac{(cb + W \tan \phi) \sec \alpha}{1 + (\tan \alpha \tan \phi)/F} = \frac{\Sigma X}{\Sigma W \sin \alpha} . \tag{2}$$

This equation must be solved for F by successive approximation.

First assume a likely value of F in the term $(\tan \alpha \tan \phi)/F$ in Equation 2. Calculate F from this equation. If this value does not agree with the one assumed, a closer approximation must be made. The calculations are most effectively set out in tabular form (Table 8.1).

In another method, proposed by Spencer, the two inter-slip forces on any strip are assumed to be parallel, but both the force and moment conditions of equilibrium must be satisfied.

Simplified method of slices

The most usual simplification, originally attributed to Krey, is to *assume* that the E forces on each strip approximately balance out. The force diagram then takes the form shown in Fig. 8.2d, and the problem is clearly statically determinate.

Compared with the more rigorous analyses, this assumption generally results in an underestimate of the factor of safety, the discrepancy being greater where the pore-pressure ratio r_u is high. This discrepancy may be a disadvantage from the economic point of view. In deciding for or against using the simplified method, we have to take into account the possible errors in the data used in the calculations, as

Table 8.1 *Bishop method – tabulation of calculations*

(a) Slice No.	(b) Sin α	(c) Height of slice gh	(d) Weight W	(e) W sin α	(f) cb + W tan φ	(g) $\dfrac{1 + (\tan \alpha \tan \phi)/F}{\sec \alpha}$	(h) $X = \dfrac{column\ (f)}{column\ (g)}$
				$\Sigma\, W \sin \alpha$			ΣX

It is usual to include the effect of pore-pressures along the probable slip circle. When this is done the effective stresses are used instead of the total stresses, and the cohesive strength becomes c' and the angle of shearing resistance ϕ'.

well as the non-uniformity of the soil and the deviation of the slip surface from a circular arc.

In the simplified analysis the procedure is as follows. The equilibrium of the strip can be represented as in Fig. 8.2d. The restraining force T tangent to the slip surface is composed of the cohesion plus the frictional resistance called into play by the normal component N of the weight of the strip corrected for pore-pressure if necessary.

As before the weight of the slice $W = \gamma hl \cos \alpha$, where α is the inclination of the element of slice surface to the horizontal. The normal component N is $W\cos \alpha = \gamma hl\cos^2 \alpha$. If there is a pore-pressure u, the effective pressure is $\gamma h\cos^2 \alpha - u$ and the effective normal force

$$N' = \gamma hl \cos^2 \alpha - ul.$$

Since the normal component passes through O, the disturbing moment is the tangential component T multiplied by the moment arm R. The total disturbing moment is therefore $R\Sigma T$.

The maximum possible restraining force on the bottom surface of the strip is $c'l + N' \tan \phi'$ and its moment about O is $R(c'l + N' \tan \phi')$. The total maximum restraining moment is the summation of these moments for all the slices and is equal to

$$R\Sigma(c'l + N' \tan \phi') = R(c'l + \tan \phi' \Sigma N').$$

The factor of safety is thus represented by

$$F = \frac{c'R\theta + \tan \phi' \Sigma N'}{\Sigma T}.$$

The weight of each slice is obtained as before, and the normal and tangential forces are found either by calculation from $N = W \cos \alpha$ and $T = W \sin \alpha$ or by drawing the triangles of forces as in Fig. 8.2d. The solution for each assumed circle can be developed in tabular form and the least factor of safety found as before.

When using the total stress analysis, the procedure is similar. The constants c_u and ϕ_u are used instead of c' and ϕ', and the total normal component on each element is $N = \gamma hl \cos^2 \alpha$.

In the more rigorous analysis it was assumed that the same factor of safety applies to the resisting forces on each slice. When the simplified method is used, neglecting the lateral forces between the slices, this assumption no longer holds good. The factor of safety for the whole bank is reckoned on the stability of the mass of soil tending to slip.

The $\phi = 0$ assumption

In the analysis of slips in saturated clay soils it is often assumed that conditions approximate to those of the undrained shear test, and therefore that the angle of shearing resistance ϕ can be taken as zero. This assumption has the advantage of simplifying the calculations. The ultimate cohesion c_u may be found from the triaxial or unconfined compression tests or, in very soft clays, from the *in situ* vane test. Investigations of a number of failures in soft clay show that the '$\phi = 0$ assumption' generally gives a reasonably correct estimate of the factor of safety. The actual slip surface, however, is usually not quite the same as that predicted by the theory.

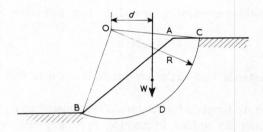

Fig. 8.3 Rotational slip ($\phi = 0$).

When the $\phi = 0$, assumption is made the simplified procedure is as follows. Refering to Fig. 8.3 the horizontal distance d of the centroid of the area *ACDBA* from the centre O is first found. This can be done by cutting out (to as large a scale as is convenient) a template of the same shape as the area and finding the centroid by suspending the template from several points, marking the vertical at each suspension. The distance d can more readily be found from the scale drawing by the use of an integrator, which gives the area of *ACDBA* and also its moment about O. The disturbing moment tending to cause slipping on the cylindrical slip surface is then Wd, where W is the weight of a unit length of the disturbed volume *ACDBA*.

The restraining moment tending to prevent the slip occurring is the product of the radius of the cylinder and the shear resistance developed along the surface *BDC*. On the $\phi = 0$ assumption, the maximum shear resistance which can be developed on any short length of the slip surface is independent of the pressure at that point and therefore of the superincumbent weight of material. This unit resistance is equal to the cohesion c_u and is assumed constant over the whole length of the slip surface L. The moment arm of the resisting force is constant for all parts of the slip surface and equal to R.

The restraining moment tending to prevent slipping can therefore reach a maximum of

$$c_u LR$$

the value of which depends only on the cohesive property of the material and the shape of the assumed slip surface. The length L is equal to $R\theta$, and the maximum restraining moment becomes

$$c_u R^2 \theta. \quad (\theta = \langle COB)$$

The disturbing moment is Wd and the factor of safety is therefore

$$F = \frac{c_u R^2 \theta}{Wd}.$$

As before, several trial circles are drawn in order to find the lowest value of F.

In the above simplified procedure, it is assumed that c_u is constant over the whole slip surface. In practice, c_u usually varies with depth, so that an average value must be taken. If it is necessary to allow for this variation, the method of slices can be used and the factors of safety found from

$$F = \frac{\Sigma c_u l}{\Sigma W \sin \alpha}.$$

When $\phi = 0$, it is found that neglecting the interslice forces has no effect on the result unless the slip surface is other than a circular arc.

Effects of cracks in cohesive soils

One of the features of rotational slips in cohesive soils is the appearance of a vertical crack running parallel to the top of the slope and at some distance from it. Such cracks can reach, as in wedge failures, a depth of $2c/\gamma$. Over this depth no shear resistance can be developed. The effect of cracking is therefore to shorten the arc *BDC* to *BDC'* (Fig. 8.4) and to decrease θ to θ'. In addition, a further

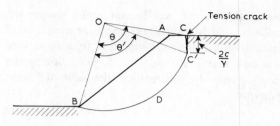

Fig. 8.4 Rotational slip with tension cracks.

disruptive force, acting against the cohesive resistance, is applied if rainwater fills the crack and exerts a hydrostatic pressure.

Taking into account of these cracks is found to have little effect on the factor of safety, and they are usually neglected.

The ϕ-circle method

This method is based on the assumption that the resultant force P on the surface of rupture is tangential to a circle of radius $R \sin \phi$ whose centre coincides with that of the surface of rupture (Fig. 8.5).

Let L be the length of the arc BD and L_1 the length of the chord of this arc. The moment about O of the cohesive force on an element of length δL is $c_m \delta LR$. Assuming uniform cohesion along the surface, the resultant of the cohesive forces on all such elements is a force $C = c_m L_1$ acting at a perpendicular distance a from O such that

$$c_m L_1 a = c_m LR$$

or

$$a = \frac{LR}{L_1} = \frac{R\theta}{2 \sin \theta/2}.$$

Having thus found the line of action of C, a line drawn through its intersection with W (the weight of the mass of soil) tangent to the ϕ-circle determines the direction of P. By constructing the force triangle the magnitude of C is found, and hence the cohesion c_m necessary for equilibrium. This procedure implies that all available frictional resistance is fully developed. The ratio of the ultimate cohesion c to the mobilized cohesion c_m may thus be termed the factor of safety with respect to cohesion.

In the definition of factor of safety previously stated it is assumed

155

that under conditions of equilibrium equal proportions of the ultimate cohesive and frictional resistances are mobilized. The radius of the ϕ-circle should therefore be reduced from $R \sin \phi$ to $R \sin \phi_1$, where ϕ_1 is such that $F \tan \phi_1 = \tan \phi$. When finding the factor of safety of a given slope by this method, the value of F must be found by trial and error.

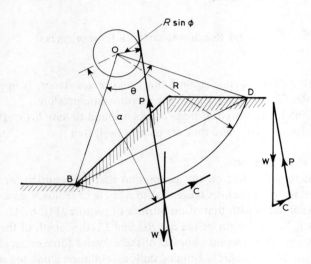

Fig. 8.5 ϕ-circle method.

The assumption that P is tangent to the circle of radius $R \sin \phi$ is not quite correct. The error is, however, small and is on the safe side. It vanishes when ϕ becomes zero, as in this case P passes through O.

In the procedure described above, no account has been taken of pore-pressures on the slip surface. To allow for this, draw a curve showing the pore pressure plotted perpendicular to the slip surface. Divide the slip surface into a number of parts, calculate the force on each and combine these vectorially to give the resultant U which must act through the centre O. The resultant is then compounded with the weight W and the reaction P to find C.

Location of the critical slip surface
Fig. 8.6 shows an example of a bank in uniform cohesive soil for which a number of slip circles have been analysed and the factor of

safety for each calculated. The 'contours' show the loci of the centres of circles for equal values of *F*, and they form long narrow loops with their axes not quite perpendicular to the slope. The position of the centre is more critically placed laterally than in the direction normal to the slope.

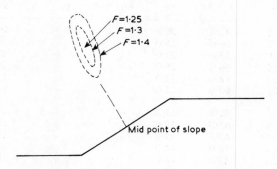

Fig. 8.6 Location of centres.

Table 8.2. shows data compiled by Taylor from a large number of investigations using an analytical solution of the ϕ-circle method on the basis of total stress. The centre of the most dangerous circular arc passing through the toe of the slope is determined by setting out angles ψ and θ as shown in Fig. 8.7a, the values of these angles for various slopes and angles of internal friction being given in Table 8.2. The sixth column of the table shows, in terms of the height of the slope *H*, the greatest depth of the rupture arc below the foot of the slope. Often the existence of a stiffer stratum of soil above this level limits the depth to which the surface of rupture can penetrate, and the critical curve is then a circular arc touching the surface of this stratum. The last column in the tables shows the *stability number.*

When *i* is less than 53° and ϕ is zero or very small, the most dangerous circle passes below the toe (Fig. 8.7b). According to Fellenius, when $\phi = 0$, the centre of this circle is situated vertically above the midpoint of the slope. Again, the depth of the arc of rupture is generally governed by the presence of rock or other firm material at a certain depth below the foot of the slope. The figures given in brackets in the table refer to the most dangerous circles passing through the toe, although a more dangerous circle may exist which passes below the toe.

157

Table 8.2 *Data for critical circles*

Slope i	Angle of Friction ϕ	Angles for setting out centre of critical circle ψ	θ	Factor n (see Fig. 8.7b)	Depth factor D	Stability No. $\dfrac{c}{F\gamma H}$
90	0	47·6	30·2	–	–	0·261
	5	50·0	28·0	–	–	0·239
	10	53·0	27·0	–	–	0·218
	15	56·0	26·0	–	–	0·199
	20	58·0	24·0	–	–	0·182
	25	60·0	22·0	–	–	0·166
75	0	41·8	51·8	–	–	0·219
	5	45·0	50·0	–	–	0·195
	10	47·5	47·0	–	–	0·173
	15	50·0	46·0	–	–	0·152
	20	53·0	44·0	–	–	0·134
	25	56·0	44·0	–	–	0·117
60	0	35·3	70·8	–	–	0·191
	5	38·5	69·0	–	–	0·162
	10	41·0	66·0	–	–	0·138
	15	44·0	63·0	–	–	0·116
	20	46·5	60·4	–	–	0·097
	25	50·0	60·0	–	–	0·079
45	0	(28·2)	(89·4)	–	(1·062)	(0·170)
	5	31·2	84·2	–	1·026	0·136
	10	34·0	79·4	–	1·006	0·108
	15	36·1	74·4	–	1·001	0·083
	20	38·0	69·0	–	–	0·062
	25	40·0	62·0	–	–	0·044
30	0	(20·0)	(106·8)	–	(1·301)	(0·156)
	5	(23·0)	(96·0)	–	(1·161)	(0·110)
	5	20·0	106·0	0·29	1·332	0·110
	10	25·0	88·0	–	1·092	0·075
	15	27·0	78·0	–	1·038	0·046
	20	28·0	62·0	–	1·003	0·025
	25	29·0	50·0	–	–	0·009
15	0	(10·6)	(121·4)	–	(2·117)	(0·145)
	5	(12·5)	(94·0)	–	(1·549)	(0·068)
	5	11·0	95·0	0·55	1·697	0·070
	10	(14·0)	(68·0)	–	(1·222)	(0·023)
	10	14·0	68·0	0·04	1·222	0·023

Figures in brackets are for most dangerous circle through the toe when a more dangerous circle exists which passes below the toe.

In practical problems it is unusual to find all the simplified conditions assumed in compiling the above data. The properties of the soil often vary at different levels, and the profile is not always a simple uniform slope bounded by level surfaces at top and bottom.

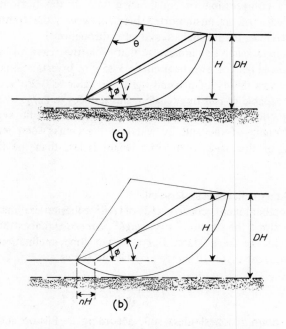

Fig. 8.7 Location of critical slip surface.

The data in Table 8.2, however, coupled with experience, will reduce the labour required to find the most dangerous slip surface in a given problem.

Spencer has published charts which locate the critical slip circle in homogeneous soil from values of the slope i, the pore-pressure ratio r_u, and the ratio $(\gamma H \tan \phi')/c'$.

Maximum height of vertical cut in cohesive soil

It is commonly stated that the maximum height at which a vertical bank in cohesive soil will stand without support is twice the depth of tension cracks, that is, $2z_c$ or $4c/\gamma \tan (45° + \phi/2)$. This is based on the assumption that in a soil having the pressure distribution shown in

159

Fig. 7.6, provided that no tension cracks develop, the resultant force down to a depth of $2z_c$ is zero. The tensile force on the upper half of this height just balances the compressive force on the lowest half, the stress distribution varying from a tension of $2c \tan (45° - \phi/2)$ at the top to a compression of equal magnitude at the bottom. On the vertical face of an unsupported bank, however, the conditions are quite different, the normal stress is zero throughout.

Slip-circle analysis indicates that for the case of $\phi = 0$ the maximum height of a vertical cut is $3.85 c/\gamma$. Investigations using the limit-analysis theory of plasticity give the value as $3.2 c/\gamma$, but, if the soil is incapable of resisting tension, this reduces to $2c/\gamma$.

In practice, the weakening effect of weathering of the vertical face and softening of the soil by water in the cracks soon reduces the stability of the bank, even if its height is less than the theoretical maximum.

Embankments in cohesionless soil

In an embankment constructed of *dry* cohesionless material the steepest slope for stability is the *angle of repose,* that is the angle of friction in the loose state. For a flatter slope, inclination i to the horizontal, the factor of safety may be defined as

$$F = \frac{\tan \phi}{\tan i}.$$

In *saturated* cohesionless soil, according to Bishop and Morgenstein, the factor of safety is reduced by the presence of pore-pressure, and is given by

$$F = \frac{\tan \phi'}{\tan i} (1 - r_u \sec^2 i).$$

It is obvious from this expression that pore-pressure effects a considerable reduction in F and a much flatter slope is therefore necessary for stability.

8.3 CRITICAL SURFACE OF RUPTURE

To the engineer the analysis of the mechanism of stability is only a step towards the design of a satisfacory bank possessing a required factor of safety. To this end it is useful to have some indication of the value of the slope which should be used in any more detailed analysis.

For geometrically similar slopes, that is, slopes with the same inclination but of different heights, the critical surface of rupture is similar in proportions, provided that the material has the same angle of shearing resistance. Thus the same drawing and the same force polygon can be applied to all such slopes merely by varying the scale. The cohesive force is the product of the unit cohesion c_m and the length of the arc of rupture, while the weight of the slipping mass is its area multiplied by its density. Since lengths are proportional to the height of the bank H, and areas to H^2, the ratio $c_m H/\gamma H^2$ or $c_m/\gamma H$ must be constant for similar slopes. Putting $c_m = c/F$, where F is the factor of safety with respect to cohesion, the quantity becomes $c/F\gamma H$. This is a dimensionless quantity and is known as the *Stability Number*. The stability number is very useful in setting up programmes for analysis by computer.

Taylor's Stability Numbers
Several authors have published charts and tables from which preliminary designs of embankments and cuttings can easily be made.

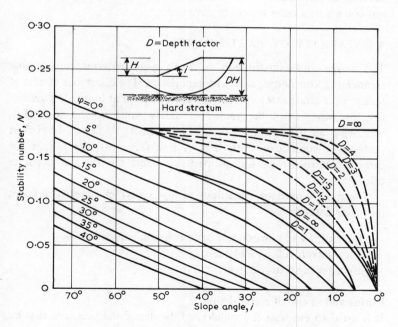

Fig. 8.8 Taylor's curves.

161

The best known are those devised by Taylor (Fig. 8.8 and Table 8.2) showing for various angles of shearing resistance the safe slopes corresponding to different values of the stability number. These curves show clearly that for cohesive material the safe slope to which the bank can be laid is dependent on its required height. The greater the height the less the stability number and the flatter must be the slope. From Table 8.2 and Fig. 8.8 it is seen that for a cohesive frictional soil there is a maximum safe angle of slope corresponding to the height of the bank. When ϕ is zero, this statement applies only if the angle of slope exceeds $53°$.

On the other hand, when cohesion is absent and the slope depends for its stability on frictional resistance alone, the safe slope is independent of the required height. When $c = 0$ the Stability Number is zero, and the maximum slope for stability is equal to the angle of shearing resistance whatever the height.

In deriving the expression for the stability number the factor of safety is applied only to the cohesion. To allow for only part of the limiting frictional resistance being developed the angle ϕ should, as before, be divided by the factor of safety. When this factor is unknown a trial value should be assumed.

8.4 STABILITY OF EARTH DAMS

The design and construction of earth dams requires considerable engineering knowledge, skill and experience. The disastrous effects of failure necessitate safety precautions as complete as humanly possible. On the other hand the factor of safety must not be too great, otherwise an uneconomic design will result. A factor of 1.5 is customary, sometimes even less. A low factor of safety calls for careful assessment of soil properties, sound theory, and accurate calculations.

The most important conditions of loading which have to be taken into consideration are:

(i)   during and immediately after construction;
(ii) reservoir full – steady seepage;
(iii) after rapid drawdown.

During and at end of construction
It is usual to estimate the stability of the dam during construction by considering effective rather than total stresses, c' and ϕ' being used.

These values are determined from samples tested in triaxial compression. Prediction of the pore-pressure ratio r_u also assists in evaluating the stability of the dam as construction proceeds.

After the placing of a layer of fill, height ΔH, the additional pressure is $\Delta\sigma_1 = \gamma\Delta H$. The pore-pressure increases from its initial value u_0 to $u_0 + \bar{B}\Delta\sigma$, where $\bar{B}$ is the pore-pressure coefficient. Hence

$$r_u = \frac{u_0 + \bar{B}\Delta\sigma_1}{\gamma\Delta H} = \frac{u_0}{\gamma\Delta H} + \bar{B}.$$

If the initial pore-pressure u_0 is zero or very small, $r_u = B$. The value of r_u for a saturated soil could therefore be unity or even greater, but it is more generally of the order of 0·5 at the densities and moisture contents at which the material is usually rolled.

This pore-pressure gradually dissipates as the fill material consolidates. Drainage blankets of permeable material are placed, in the body of the embankment, particularly if it is a water-retaining structure, in order to accelerate the reduction of pore-pressure as well as giving indications of dangerous rises in pore-pressure as construction proceeds.

In large dams, piezometers are generally built in at various points, enabling a continuous record of pore-pressures to be made. The results provide a valuable check on the validity of the assumptions made in the stability calculations.

Rapid drawdown

If the water in a reservoir is drawn off rapidly (a few days, or even weeks, may be 'rapid' in the soil mechanics sense) there is a temporary reduction in the factor of safety, until equilibrium is eventually reached under the new conditions.

Rapid drawdown has two closely related effects:

(i) the weight of the soil mass tending to slip is increased, since its density changes from submerged to bulk;

(ii) the shearing resistance on the slip surface is reduced on account of the slow dissipation of the residual pore pressure, the presence of which reduces the effective stress.

Referring to Fig. 8.1, before drawdown the pore-pressure at the point A at depth to below the water surface was

$$u_0 = \gamma_w(h + h_1)$$

(h_1 may be positive or negative according to the initial flow pattern). After rapid drawdown the pore-pressure becomes

$$u = u_0 + \Delta u = u_0 + \bar{B}\Delta\sigma_1.$$

The change $\Delta\sigma_1$ in vertical pressure is the water load removed, that is

$$\Delta\sigma_1 = -\gamma_w(h - h_2)$$

where h_2 is the height of the final water level above A.

$$u = \gamma_w[h + h_1 - \bar{B}(h - h_2)]$$
$$= \gamma_w[h(1 - \bar{B}) + h_1 + \bar{B}h_2].$$

From this it is seen that by reducing B, u is increased. Since B is generally greater than unity, it is usual to take $B = 1$ as the worst case, giving

$$u = \gamma_w(h_1 + h_2).$$

If the dam has a rock-fill face, allowance must be made for the water which drains out rapidly, leaving the weight of the rock-fill itself still acting.

Taylor suggests an approximate solution to the rapid drawdown problem, making use of the Stability Number. In the equation for the Stability Number, $N = c/F\gamma H$, γ is taken as the bulk density of the soil, and if $\phi > 0$ the friction angle to be used in the tables or chart is reduced to

$$\frac{\gamma - \gamma_w}{\gamma} \phi.$$

Reservoir full – steady seepage

When equilibrium is attained, a condition of steady seepage is set up. This is of special importance for the stability of the downstream part of the dam. It was shown in Fig. 8.1 how the pore-pressure at any point can be estimated from a flow net. In this way the forces due to pore-pressure can be calculated and included in the stability analysis for a trial slip surface.

8.5 REMEDIAL AND PREVENTIVE MEASURES

The methods described in this chapter are frequently applied to the investigation of slips which have occurred, with the object of supplying the engineer with data from which he can prescribe suitable remedial measures. The surface on which the slip has taken place is carefully surveyed by digging trenches or driving headings through the displaced material. In deep-seated slips where excavation is impracticable, the condition of samples extracted from boreholes often gives an indication of the depth of the surface of rupture. The slip curve is plotted to scale and the stability is then investigated, the Swedish method being the most suitable for this purpose. It is often possible to simplify the working by finding a circular arc which approximates closely to the plotted slip curve. Sometimes a better approximation is obtained by assuming the curve to consist of two or more arcs of different radii.

A method of determining the location of the slip surface in a borehole has been used successfully. On a moving slope several boreholes are driven and into each is inserted an alkthene tube, the tops being suitably covered and concealed. As further slip takes place, the tube bends and the insertion of a thin cylindrical plumb-bob allows of the location of the kink in the tube. Whereas the plumb-bob could, originally, pass to the botton of the bore, after slip takes place it can penetrate only to the level of the slip surface. Plumb-bobs of different lengths can be used for more accurate location, the shorter bobs being the least likely to be obstructed.

Remedial works are carried out with the object of stabilizing the portions of the slopes where slips have occurred or where they are likely to occur. This can be achieved:

(i) by providing some external support;
(ii) by removing some of the weight tending to cause slipping;
(iii) by increasing the strength of the soil within which the slip surface occurs.

External support

For small slips success is likely to be achieved by holding back the toe of the slope. This can be done by piling through the toe, by building a retaining wall along the toe, or by heavy external loading on the toe. Loading the toe is usually considered as a temporary expedient, but may be sufficient to retain the slip until permanent measures are adopted. The use of piles or a retaining wall is satisfactory only when their proportions are sufficiently large compared with the height of the slope in question and the foundations extend below the slip surface. When the height of the wall or the length of the piles represents only a small proportion of the height of the embankment or cutting, little support can be expected.

Removal of material
If the material which is likely to slip is reduced in quantity at the top of the slope, the moment Wd is reduced because of a reduction both in W and in d. This type of remedy can be applied as shown in Fig. 8.10.

The most obvious method of reducing the weight of soil tending to cause a circular slip is by flattening the slope. By this means, except for extremely weak material, the factor of safety can be increased to a satisfactory figure. The method is, however, uneconomical in that a much wider strip of land is required for any embankment or cutting than is necessary with a steeper slope (Fig. 8.9a).

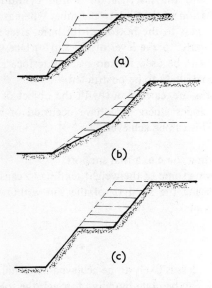

Fig. 8.9 Remedial methods.

The sloping surface of a slipped bank sometimes passes through different strata, with a stronger and more frictional material on top. In such circumstances the weaker material, which is more likely to slip, can be trimmed to a flatter slope, while the frictional material is maintained at the steeper and more economical angle (Fig. 8.9b).

Finally, the weight of the upper portion of the bank can be reduced by the insertion of one or more berms. In Fig. 8.9c a berm at the central height is shown, and, as in the other diagrams in Fig. 8.9, the shaded area represents the volume of soil removed. This method again involves the acquisition of a wider strip of land, but is a method much used, especially for earthen dams. It is also used for slopes where the horizontal distance between bottom and top of the bank is restricted, and a flatter slope is impossible. The hatched portion is carted away.

Drainage
Most slips occur in cohesive material and take place because the soil has become weakened by an increase in moisture content. A reduction in the moisture content, then, results in a rapid and substantial improvement in stress-resisting properties.

Two problems arise:

(i) the removal of water already in the fissurs and voids of the soil;
(ii) the prevention of further softening by the removal of surface water before it has time to penetrate.

Surface and subsurface drains whose primary function is to remove water rapidly may also fulfil the subsidiary function of supporting the slope against further slip. If slip has already taken place, one of the best remedial measures is to build counterfort drains running at right angles to the length of the bank and extending beyond the slip plane. These drains, which take the form of trenches filled with hardcore or brick rubble, break up and support the mass which is likely to slip and also remove rapidly any water in the soil.

Such counterfort drains should be assisted in their function by a herringbone or chevron system of surface-water drains laid on the surface of the slope. Surface drains have no effect on the moisture content of the main mass of the soil, but prevent the entry of surface water which might cause further softening. The strengthening of the surface may also be accomplished by such obvious remedies as the provision of drainage through layers of ashes, or the turfing of slopes to bind the soil and provide transpiration and shedding of moisture.

Much can also be done to improve the qualities of the soil by the laying of interceptors along the top of the slope. Surface water from higher levels is thus trapped and removed before it reaches the vulnerable material. This method is particularly valuable where a permeable soil overlies a cohesive stratum, but such interceptor drains must usually be deep.

When an earth dam is constructed of relatively impermeable material, dangerous pore-pressures may be set up during the construction period. A break in the process of tipping gives time for pore-pressures to be dissipated. The values of these pressures must be continuously recorded from piezometers. Another method of reducing the pore-pressure is to accelerate its dissipation by means of horizontal blankets of permeable material, sometimes with vertical sand drains connecting them. The vertical spacing of the blankets and the pattern of the sand drains depend on the magnitude of the pore-pressures, the rate at which they have to be dissipated, and the consolidation properties of the fill material.

8.6 APPLICATION OF THE WORK OF CHAPTER 8

The studies in Chapter 8 are concerned with the analysis of stability of slopes and earth dams which have slipped or are showing signs of instability. They also apply to the design of such earth structures before they are constructed. The problems presented should be worked over again. A good deal of the work is graphical, and a skill at drawing flow nets or in estimating the most likely type of slip surface can expedite the work.

Application 8A
It is required to find the factor of safety for a bank, 13 m high standing at a slope of 45°. The density of the soil is 2·08 Mg m⁻³ the cohesion is 28 kN m⁻² and φ = 20°.

(i) Make a sketch of the slope with dimensions and properties of the soil.

(ii) Enter Table 8.2 with values of slope and angle of shearing resistance and find the stability number to be 0.062.

(iii) Thus

$$F = \frac{c}{N\gamma H} = \frac{28}{0 \cdot 062(2 \cdot 08 \times 9 \cdot 81)13} = 1 \cdot 7.$$

(iv) This is the factor of safety with respect to cohesion only. The true value of the factor of safety will be somewhat less than this. If a trial value of $F = 1 \cdot 5$ is taken, the reduced friction angle is $20/1 \cdot 5$ or $13 \cdot 3°$. For this angle, by interpolation in the Table 8.2, the Stability Number is $0 \cdot 091$ which, in turn, gives $F = 1 \cdot 16$. The correct value of F lies between these values and further iteration shows it to be about $1 \cdot 3$.

Bishop and Morgenstern* have published a set of stability curves which take into account the pore-pressure ratio. For a given slope, F is found to vary linearly with r_u over a range of r_u from 0 to $0 \cdot 7$, and may thus be expressed in the form $F = m - n r_u$, where m and n are *stability constants*. The authors give tables and charts showing values of m and n for various angles of slope and values of ϕ' and r_u.

Spencer gives charts showing the stability number N plotted against the slope i for values of $r_u = 0, 0 \cdot 25$, and $0 \cdot 50$ for a series of values of ϕ'.

* *Geotechnique* Dec. 1960

Application 8B

A partially submerged embankment, $9 \cdot 2$ m high, is to be constructed at a slope of $1\frac{1}{2}$ to 1 (Fig. 8.10). The soil is saturated. The existing pore-pressures, found from a flow-net investigation, correspond to a phreatic surface AB. For the single trial slip surface shown, find the factor of safety of the slope for steady-seepage conditions. The properties of the two layers of soil are shown in Fig. 8.10.

(i) Take $\phi = 10°$ for slice No. 6 and $20°$ for the rest. Then the maximum resistance forces developed by cohesion and friction are

$$C + \Sigma N' \tan \phi$$

where C is the total cohesive force on arcs CD and DE, and N' is the effective normal stress on each slice of the slip surface – the total normal stress less the pore-pressure.

(ii) Maximum cohesive resisting force

$$C_1 \text{ arc } CD + C_2 \text{ arc } DE$$

$$= 12 \times 3 \cdot 5 + 29 \times 18 \cdot 8$$

$$= 587 \text{ kN.}$$

(iii) Weight of slices and pore-pressures on the slip surface at the centre line of each slice:

For a partially submerged slope, allowance must be made for the hydrostatic pressure acting on the submerged part of the slope. This pressure increases stability. For all submerged slices (e.g. slice No. 1)

Weight of slice: $W = \gamma b h_1 + \gamma_w b h_2$
Pore-water force: $U = \gamma_w (h_1 + h_2)l$
where l is the length of the chord of the surface of the slice.

For all unsubmerged slices (e.g. slice No. 3)

$$W = \gamma b (h_3 + h_4)$$

$$U = \gamma_w h_3 l.$$

(iv) Taking the summations from (ii) above and from Table 8.3, Factor of safety

$$F = \frac{C + \Sigma N' \tan \phi}{\Sigma T}$$

$$= \tfrac{1}{532}(587 + 548 \tan 20° + 68 \tan 10°)$$

$$= 1.5.$$

This factor of safety applies only to this trial slip circle. Other possible slip circles must be tested before the least F. of S. can be found.

Table 8.3 *Application 8B*

Strip No.	Weight	Components		Pore-water force	Effective normal force
		Tangential	Normal		
	W	T	N	U	N' = N − U
1	133	−18	130	101	29
2	201	14	199	102	97
3	290	76	278	130	148
4	333	150	295	137	158
5	305	198	237	121	116
6	137	112	78	10	68
		532			548 + 68

All forces are in kN.

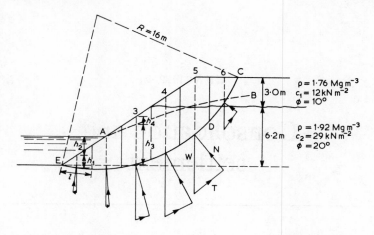

Fig. 8.10 Partially submerged embankment (Application 8B).

CHAPTER 9

Consolidation and settlement

When a compressive load is applied to the surface of the soil the resulting deformation is elastic, plastic, compressive or a combination of these.

Elastic deformation causes lateral bulging with little change of porosity, and the material recovers on removal of the load. When treading on fairly dry or spongy peak soil, the foot sinks in a little, but if it is lifted quickly the footmark practically disappears. This is mainly an elastic deformation.

Plastic deformation is due to the lateral flow of the soil under the action of pressure, with negligible rebound after removal of the load. *Plasticity* is the property which enables the material to undergo considerable deformation without cracking or crumbling. All clays have this property to a greater or less degree provided that the moisture content is not below the plastic limit. In soft clay or mud we observe the soil squeezing outwards under the sole of the foot, indicating plastic flow.

Compressive deformation occurs when the particles are wedged closer together by pressure, reducing the pore space by the expulsion of water and air. This property of soil is known as *compressibility*. On medium clay or on an ordinary garden flower bed, the foot sinks in and leaves a footmark, the soil underneath having been compressed or consolidated.

The compression of soil under steady pressure, such as the weight imposed by a structure or earth filling, is known as *consolidation* and is due to the expulsion of water from the pores. When the compression is effected on unsaturated soils by mechanical means such as rolling or tamping, it is termed compaction, and air, rather than water, is driven out.

In problems of consolidation, both the magnitude of the settlement and the rate at which it occurs are of practical importance. The magnitude depends on the compressibility of the soil, the effect being defined by the *coefficient of compressibility*. The *rate of consolidation* depends on permeability as well as compressibility, and it is convenient to combine the effects of these two soil properties in a factor known as the *coefficient of consolidation*. In the usual theory of consolidation it is assumed that drainage is in the vertical direction only.

In sands, consolidation may generally be considered to keep pace with construction, and the after-effects are therefore much smaller than in fine-grained soils. The subject matter of this chapter is therefore primarily concerned with the compressibility of silts and clays.

9.1 COMPRESSIBILITY

Clay, as we find it in the ground, has undergone a natural process of consolidation, having been originally deposited in water and then gradually compressed by the weight of the material deposited above. The soil is said to be fully or partially consolidated, according to whether or not a state of equilibrium has been reached. Some clays are *over-consolidated:* they have been compressed under a super-imposed load, such as that of the ice-sheets of the Pleistocene period, or have consolidated because of free-draining conditions, as in some lodgement tills. The compressibility and consolidation properties of the soil are determined from the oedometer test, which is described more fully in a later Chapter.

In this test a sample of soil is confined in a flat mould between two porous discs which allow free access of water to or from the soil. A compressive load is applied and allowed to act until equilibrium is reached. Readings of the reduction in thickness of the sample are taken at appropriate time intervals. Fig. 9.1 shows a typical time/compression curve for a sample of normally consolidated clay. It must be noted, that, in this test, drainage is in the vertical direction only.

After the time/compression readings for the first load are recorded, an increment of pressure is applied and the readings for another time/compression curve are observed. The process is repeated for several further increments of pressure, as shown diagrammatically in Fig. 9.1b. In this diagram the curve AB is the time/compression curve obtained under the first steady pressure p_1. The pressure is then increased to p_2 and the curve BC is obtained, and so on. Finally the pressure is released and the sample allowed to take up water and expand. After the expansion is complete, the final moisture content of the soil is determined.

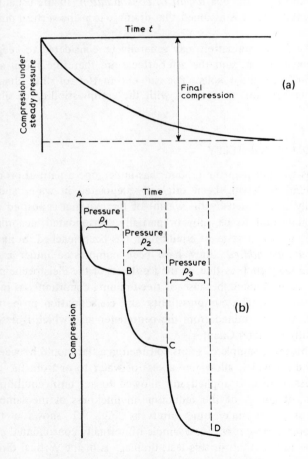

Fig. 9.1 Time/compression curve under successive increments of pressure.

Relation between pressure and void ratio

Knowing the final moisture content m at the end of a consolidation test on saturated soil, the final void ratio is found from the equation $e = mG_s$. From the changes of thickness of the sample which are proportional to the volumetric changes, from stage to stage in the test, we can estimate the void ratio at the end of each of these stages.

The *pressure/void ratio curve* can now be plotted. In Fig. 9.2. *AB* is a typical pressure/void ratio curve for the compression of a sample of *normally-consolidated* clay which has suffered no extra pressure, but only the effect of its own weight. If, at a point such as *D*, the pressure is reduced, the soil expands and takes up water, but the expansion curve is much flatter than the compression curve. Consolidation may thus be regarded as consisting of two parts:

(i) *reversible* – the reduction in the thickness of the moisture films separating the particles;

(ii) *irreversible* – a change in the orientation of the particles.

Pressure/void-ratio curve

In the course of the geological history of *overconsolidated* soils (which are soils which have been compressed by externally-applied load), some of the overlying material has been eroded and the consolidating pressure released. The effect of compression, expansion, and re-compression is shown in Fig. 9.2, where *AD* is the original compression curve, and the full line *DE* represents the expansion which takes place on release of overburden pressure. If the soil is re-compressed by fresh deposits, the new compression curve takes the form of the dotted line from *E*. If there is a further increase of pressure beyond *D*, the curve becomes a continuation of the original curve *ADB*.

In the oedometer test, an undisturbed sample of overconsolidated clay will give a flattened or convex-upward compression curve of shape similar to *EDB* in the diagram.

Compression index. The pressure/void ratio curve for cohesive soil, *AB* in Fig. 9.2, corresponding to natural consolidation was termed, by Terzaghi, the *virgin consolidation curve*. Between the liquid and plastic limits the curve is of the exponential type and can be expressed by the empirical equation

$$e = e_1 - C_c \log_{10} p$$

175

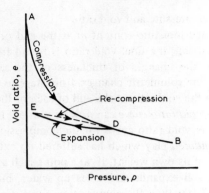

Fig. 9.2 Effect of re-compression.

where C_c is the *compression index* and e_1 is the void ratio at unit pressure. The curve of e plotted on a base of $\log p$ is a straight line.
The slope has a value of C_c as shown in Fig. 9.3. Similarly the expansion curve plotted on a logarithmic base is approximately a straight line the slope of which gives the *expansion* or *swelling index, C_e*.

For overconsolidated clay the $e/\log p$ curve is of the form $E'D'B'$. Casagrande suggested an empirical construction (Fig. 9.4) to estimate from this curve the probable pre-consolidation pressure, which is often of interest in tracing the geological history of the soil. The procedure is to estimate the point D of maximum curvature of the

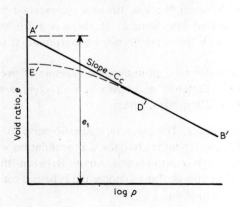

Fig. 9.3 Determination of compression index.

e/log p graph, draw a tangent at this point, and bisect the angle between the tangent and a horizontal line through D. The point E where the bisector cuts the straight part of the graph represents approximately the pre-consolidation pressure, which is another term used, or *maximum past consolidation pressure*. The overconsolidation ratio is the ratio of the maximum past consolidation pressure to the present pressure on the soil.

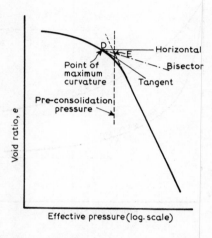

Fig. 9.4 Determination of pre-consolidation pressure.

Relation between coefficient of compressibility and compression index. In normally consolidated soil the relation between e and p is

$$e = e_1 - C_c \log_{10} p$$
$$= e_1 - 0.434 C_c \log p.$$

Differentiating with respect to p,

$$\frac{de}{dp} = -\frac{0.434 C_c}{p}$$

and

$$m_v = -\frac{1}{1 + e} \frac{de}{dp} = \frac{0.434 C_c}{(1 + e)p}.$$

177

Effect of remoulding

Fig. 9.5 shows pressure/void-ratio curves for two samples of clay, consolidated to equilibrium, and starting from the same void ratio. The upper curve refers to an undisturbed sample and the other to a remoulded sample. Remoulded samples may range from a slurry made

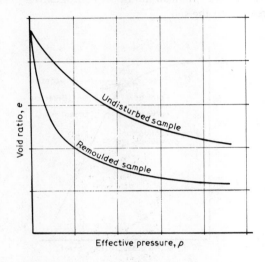

Fig. 9.5 Effect of remoulding on p/e curve.

from clay at the liquid limit, to a test sample broken up and squeezed and moulded into a shape suitable for testing. The curve is similar in shape to the virgin consolidation curve but lies below it. This is because the natural complex structure of the clay, as originally deposited, cannot be reproduced artificially.

9.2 CONSOLIDATION IN ONE DIRECTION

When an increment of pressure is applied to a sample of soil, the pressure is at first carried entirely by the pore-water. A hydraulic gradient is set up and water commences to drain away both upwards and downwards according to Darcy's law. As the water escapes, more and more of the pressure is transferred to the soil particles, the head is reduced, and the rate of flow decreases. Eventually final consolidation is reached with a much decreased void ratio, and the additional applied pressure is then carried entirely by the soil particles.

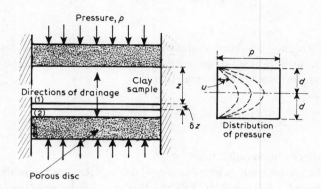

Fig. 9.6 Consolidation test in oedometer.

The relationship between pressure and void ratio for a particular soil (the p/e curve discussed above) can be used to determine

(i) the *coefficient of compressibility* (which leads to estimation of total settlement);
(ii) the *coefficient of consolidation* (which allows calculation of the rate at which settlement takes place).

Of these, the *coefficient of compressibility* (m_v) is the more frequently used, and is of primary importance.

Coefficient of compressibility

The *coefficient of compressibility*, or *volume compressibility*, is the *decrease in unit volume for each unit increase in pressure*. The decrease in volume of the sample is caused by reduction in the void ratio with pressure. In the oedometer, that reduction is proportional to the decrease in thickness. The *coefficient of compressibility* can, therefore, be expressed in terms of thickness. For a constant cross-sectional area, the coefficient of compressibility is the *decrease in unit thickness for each unit increase in pressure*.

179

$$m_{\mathrm{v}} = \frac{\text{change in unit thickness}}{\text{change of pressure}} = \frac{\delta h}{h} \, \delta p$$

$$m_{\mathrm{v}} = \frac{1}{h} \frac{\delta h}{\delta p} .$$

When m_{v} has been evaluated from the consolidation test, the change in thickness, or the settlement of a layer of clay can be found by transforming the above equation.

$$\delta h = m_{\mathrm{v}} h \delta p$$

The theory of consolidation

In the oedometer, where drainage takes place both upwards and downwards, d is equal to half the thickness of the soil sample (Fig. 9.7).

Referring to Fig. 9.6, consider an elementary layer of thickness δz and unit cross-sectional area. The hydraulic pressure gradient $= \partial u / \partial z$.

Change of $\dfrac{\partial u}{\partial z}$ in length $\delta z = \dfrac{\partial^2 u}{\partial z^2} \delta z$.

Hydraulic gradient at depth $z = \dfrac{1}{\gamma_{\mathrm{w}}} \dfrac{\partial u}{\partial z} = i_1$.

Hydraulic gradient at depth $z + \delta z = \dfrac{1}{\gamma_{\mathrm{w}}} \left(\dfrac{\partial u}{\partial z} + \dfrac{\partial^2 u}{\partial z^2} \delta z \right) = i_2$.

By Darcy's law, velocity at section (1) $= ki_1 =$ rate of inflow per unit area.

Velocity at section (2) $= ki_2 =$ rate of outflow per unit area.

$$\text{Water lost per minute} = k(i_2 - i_1) = \frac{k}{\gamma_{\mathrm{w}}} \frac{\partial^2 u}{\partial z^2} \delta z.$$

This is equal to the rate of reduction of the volume of the element.

$$\text{The volumetric strain} = - m_{\mathrm{v}} \delta (p - u).$$

(The negative sign denotes decrease in volume with increase in pressure.)

Therefore rate of change of volume =

$$-m_v \frac{\partial(p-u)}{\partial t} \delta z = m_v \frac{\partial u}{\partial t} \delta z.$$

since p is constant and $\dfrac{\partial p}{\partial t} = 0$.

$$\frac{k}{\gamma_w} \frac{\partial^2 u}{\partial z^2} \delta z = m_v \frac{\partial u}{\partial t} \delta z.$$

Put

$$\frac{k}{\gamma_w m_v} = c_v \text{ (coefficient of consolidation)}.$$

Then

$$\frac{c_v \partial^2 u}{\partial z^2} = \frac{\partial u}{\partial t}.$$

This is the basic differential equation of consolidation, which includes the effects of permeability (k), compressibility (m_v) and rate of consolidation (c_v).

Degree of consolidation, U_v represents the ratio of the amount of consolidation effected, to the total ultimate consolidation. Thus, half the ultimate consolidation represents a degree of consolidation of 0·5. Solutions have been found for certain conditions of pressure distribution and drainage in the form of equations connecting the degree of consolidation U_v with a dimensionless time factor, T_v.

For the particular case of uniform pressure distribution throughout the depth of the compressible stratum the solution is:

$$\left[\epsilon^{-\frac{\pi^2 T_v}{4}} + \frac{1}{9} \epsilon^{-\frac{9\pi^2 T_v}{4}} + \frac{1}{25} \epsilon^{-\frac{25\pi^2 T_v}{4}} + \ldots \right]$$

Expressions of a similar character have been derived for certain cases of non-uniform pressure distribution.

Degree of consolidation and time factor

As the time factor T_v involves both the length of the drainage path d and the coefficient of consolidation c_v the same time factor/consolidation curve will apply to any practical problems in which the pressure distribution and drainage conditions are the same. Table 9.2 shows corresponding values of T_v and U_v for several cases of pressure distribution.

Table 9.1 *Relation between degree of consolidation and time factor*

Types of pressure distribution
(one-way flow)

Top (permeable)	(1)	(2)	(3)

Bottom (impermeable)

Time factor T_v

Degree of consolidation	Condition (1)	Condition (2)	Condition (3)
0·1	0·008	0·047	0·003
0·2	0·031	0·100	0·009
0·3	0·071	0·158	0·024
0·4	0·126	0·221	0·048
0·5	0·197	0·294	0·092
0·6	0·287	0·383	0·160
0·7	0·403	0·500	0·271
0·8	0·567	0·665	0·440
0·9	0·848	0·940	0·720

For two-way drainage, Condition (1) is used for all linear distributions of pressure, and d is taken as half the thickness of the layer.

The rate at which the degree of consolidation (U_v) increases depends on:

(i) the distribution of effective pressure across the thickness of the stratum;

(ii) the length of the drainage path;

(iii) whether water can escape from one or both surfaces of the stratum;

(iv) the coefficient of compressibility of the soil;
 (v) the coefficient of permeability of the soil.

At this stage, two significant relationships should be noted

$$c_v = \frac{k}{\gamma_w m_v}$$

and

$$c_v = \frac{T_v d^2}{t}$$

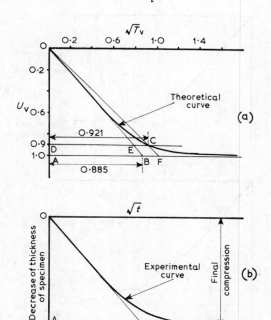

Fig. 9.7 Rate-of-consolidation curves.

Coefficient of consolidation

The rate of consolidation of a loaded clay stratum is sometimes of importance. The many years which a thick, stiff clay layer requires to reach complete consolidation may not be of significance. But, if the affected stratum is soft and thin, the rate of consolidation will be more rapid, and may have an influence on decisions required during construction of the structure. The values of rates of settlement, predicted from calculated *coefficients of consolidation,* do not often agree closely with those observed in practice, but at least a measure of the order of the rate of settlement can be evaluated.

The first step in developing a *coefficient of consolidation* is to add to the observations in the consolidation test, a note of the times taken for each increment of load to reach an equilibrium, and the sample to

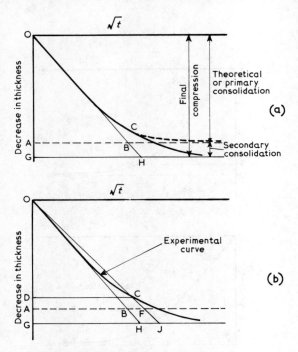

Fig. 9.8 Alternative method for coefficient of consolidation.

attain a stable thickness under each increment. The curve of time against change in thickness of sample $(t/\delta h)$ is found to agree reasonably well with the curve of the related dimensionless ratio (T_v/U_v) obtained from solution of the theoretical equation.

A variation of the curve T_v/U_v to $\sqrt{T_v}/U_v$ and the curve of $t/\delta h$ to $\sqrt{t}/\delta h$ (Figs. 9.7a and b) shows that the curves of theory and experiment not only match, but show (to 50 to 60 per cent of consolidation) a straight portion. The equation of this straight portion is found to be

$$U_v = \frac{2\sqrt{T_v}}{\sqrt{\pi}}.$$

When $U_v = 1 \cdot 0$ (100 per cent consolidation), $\sqrt{T_v} = 0 \cdot 5\sqrt{\pi}$ T_v is thus $\pi/4$ or 0.785.

Thus, at $U_v = 1 \cdot 0$, the expression

$$c_v = \frac{T_v d^2}{t} \text{ becomes } \frac{\pi d^2}{4t_1}.$$

The simplest development of the *coefficient of consolidation* evolves from making the simplifying assumption that the straight line relationship extends to the maximum consolidation ($U_v = 1$) under any load. In the theoretical curve, $\sqrt{T_v}$ will be 0·885 (Fig. 9.7a), and in the experimental curve $\sqrt{t}$ will be called $\sqrt{t_1}$ which can be measured in the consolidation test.

At total consolidation under the load, when $U_v = 1$ and $\sqrt{T_v} = 0 \cdot 885$ or $0 \cdot 5\sqrt{\pi}$ then

$$c_v = \frac{\pi d^2}{4t_1}.$$

For a particular stratum, the drainage path d is known; this 'stratum' in the consolidation test is the thickness of the sample. The value of t_1 is found by projecting the straight-line condition through to 100 per cent consolidation, and measuring off $\sqrt{t_1}$. The value of c_v can then be calculated.

Coefficient of consolidation using log time. A method (proposed by Casagrande) of obtaining the coefficient is shown developed numerically in Application 9C at the end of the chapter.

Divergence of theoretical and experimental consolidation

The value of c_v obtained in the last section was based on a simplification of what actually occurs, but allowed of the comparison and fitting to one pattern, of the theoretical $U_v/\sqrt{T_v}$ curve, and the experimental $\delta h/\sqrt{t}$ curve. The calculation was based on the assumption that a straight line relationship existed in both these curves up to 100 per cent consolidation.

185

In practice, although the theoretical and experimental curves agree closely, even to a degree of primary consolidation of $U_v = 0.9$, at least a third of the curve deviates from the straight line used in the earlier derivation, which takes no account of the slowing down of settlement in the final stages.

In this alternative method of determining the *coefficient of consolidation* instead of using the fictitious point B (Fig. 9.7) at 100 per cent consolidation, the point C (on both curves) is used at 90 per cent primary consolidation (Fig. 9.8). Up to this point the theoretical and experimental curves agree reasonably well, but as consolidation continues, the experimental curve deviates from the theoretical curve of primary consolidation. This *secondary compression* is considered in the next section; for the present, only that part of the two curves down to $U_v = 0.9$ (90 per cent primary consolidation) is considered. The problem is to find the value of $\sqrt{t}$ for the point C representing not 90 per cent of the measured experimental consolidation, but 90 per cent of the theoretical primary consolidation. This is done by linking the curves $U_v/\sqrt{T_v}$ and $\delta h/\sqrt{t}$ in the following steps:

Keep Figs. 9.7a and 9.8b under review, remembering that the only curve drawn after the consolidation test is the dark 'experimental curve'. The position of point C is not known and must be determined from comparison with the theory of primary consolidation:

(i) Draw the line OEB as before through the straight portion of the experimental $\delta h/t$ curve (Fig. 9.8b).

(ii) Find the value of T_v for 0·9 degree of consolidation (U_v). From Table 9.2 this is 0·848, and thus $\sqrt{T_v}$ is 0·921. This length is DC in Fig. 9.7a and 9.8b.

(iii) The line OE represents the straight-line relationship, and the line OC has a flatter slope by the ratio DC/DE or 0·921/0·798 ($0.9\sqrt{\pi}/2 = 0.798$), which is 1·15.

(iv) OC is found by extending OB to some point H and making $GJ = 1.15\,GH$.

(v) Join J to O. Where this line passes through the experimental curve is the point C at 90 per cent primary consolidation – the limit of the agreement of theory and experiment.

(vi) Measure DC on the $\sqrt{t}$ scale and convert to t. This is the value of the time required to reach 90 per cent primary consolidation. At that point on the theoretical curve $T_v = 0.848$ ($\sqrt{T_v} = 0.921$).

The expression

$$c_v = \frac{T_v d^2}{t}$$

$$= 0.848 d^2 / t$$

and can thus be evaluated for a point on the primary consolidation curve. The value of c_v will be similar to that obtained by the simpler convention, but perhaps closer to the true value. This still takes no account of secondary compression which must also be considered.

Secondary compression or consolidation

In Figs. 9.8 a and b, it can be seen that the experimental curve of compression or consolidation does not become asymptotic to the line representing 100 per cent primary consolidation as shown in Fig. 9.7b. Instead, it continues downward below the lines *AB* showing a continuing or *secondary compression*.

This secondary effect cannot be calculated by the theory of consolidation, which deals only with deformation due to the dissipation of excess pore-pressure under load. In the consolidation test, secondary compression takes place after the dissipation of pore-pressure and is due to a slow viscous change. There is a very gradual decrease in void ratio without moisture movement.

Tests on London clay show that, for that material, the value of the coefficient of consolidation (which controls rate of settlement) seems to increase by 10 or 15 per cent as the effect of secondary consolidation. This figure cannot be accepted for all clays, and indeed the sequence of settlement – 'primary' followed by 'secondary' – may also be misleading. In the thin sample of the consolidation test these two may follow in sequence, but in the long years of settlement of a thick stratum, it is possible that the dissipation of pore pressure can take place simultaneously with the time-dependent 'secondary' consolidation, thus masking the effect of the latter.

The first part of the experimental consolidation curve takes place very quickly, and may be even more rapid if the soil is non-saturated. At the other end of the curve, the effects of secondary consolidation mask the point at which 100 per cent consolidation occurs. A method which allows of the locating of the theoretical zero and 100 per cent primary consolidation was developed by Casagrande and takes into account the slope of the secondary consolidation portion of the curve. See Application 9C.

9.3 THREE-DIMENSIONAL CONSOLIDATION

In the oedometer test, drainage is in the vertical direction only and lateral strain is prevented. These conditions are assumed in the conventional theory of unidirectional consolidation. Such conditions do not always apply in practice, but are approximately satisfied in two cases. These are (i) a relatively thin layer of clay lying between incompressible strata and (ii) a large loaded area of which the horizontal extent is great compared with the thickness of the underlying clay. In the latter case the lateral strains are negligible except near the outer edges of the loaded area.

In other cases, when the one-dimensional theory is used to predict settlement, two sources of error may be present:

(i) There is generally some lateral drainage, though this is usually small since the horizontal drainage paths are likely to be much longer than the vertical. On the other hand the clay may be anisotropic, that is the horizontal permeability may be considerably greater than the vertical.

(ii) If the thickness of the compressible soil is not small compared with the dimensions of the loaded area, the lateral strains, caused by the loading, affect the pore pressures which govern the drainage flow, thus modifying both the magnitude and the rate of consolidation.

It might be expected that a set of differential equations of the type

$$\frac{c_v \partial u}{\partial z^2} = \frac{\partial u}{\partial t}$$

could be formed to represent the case of three-dimensional consolidation, similar to those for the flow of heat. The problem is, however, not quite so simple. In the methods described for determining the stress distribution in the soil, elastic conditions are assumed, that is the stress/strain relation is taken as linear and the strains are assumed to be relatively small. For soils these assumptions are not fully justified, and in the consolidation problem there is a further complication, the variation in the strains due to the changes in the elastic constants which occur as consolidation proceeds. This in turn alters the stress distribution and therefore modifies the pore-pressures.

Several theories have been developed in which these various factors are taken into account, but they are difficult to apply to practical problems.

Modification of one-dimensional theory

A procedure by which the conventional one-dimensional theory of consolidation can be modified for the more general case has been suggested by Skempton and Bjerrum. The settlement estimated by the ordinary method is multiplied by a factor, μ, which depends on (i) the pore-pressure coefficients as found from triaxial tests and (ii) the ratio of the thickness of the clay stratum to the breadth of the foundation.

In the oedometer test, where later strain is prevented, $\Delta u = \Delta \sigma$, and is independent of the lateral pressure. In the soil below a foundation, however, when an increment of vertical pressure $\Delta \sigma$, is applied, the change in pore-pressure is given by

$$\Delta u = B[\Delta \sigma_3 + A(\Delta \sigma_1 - \Delta \sigma_3)].$$

For saturated clay B may be taken as unity. Therefore

$$\Delta u = \Delta \sigma_1 \left[A + \frac{\Delta \sigma_3}{\Delta \sigma_1} (1 - A) \right].$$

To express this in a simpler form put $\Delta \sigma_3 / \Delta \sigma_1 = \alpha$ and write the factor $A + \alpha (1 - A)$ as μ, so that $\Delta u = \mu \Delta \sigma_1$.

The ratio α depends on the geometry of the problem, that is

(i) the shape of the loaded area;
(ii) the ratio h/b (depth of compressible stratum to breadth of foundation).

Table 9.2 gives values of α for circular and strip footings for various values of the ratio h/b. From these, if A is found from a triaxial test, or if a suitable value can be assumed, the factor μ can be calculated.

Let s_{oed} be the settlement, calculated by the conventional theory, produced by an applied pressure p. Here

$$\Delta \sigma_1 = p = \Delta u$$

Therefore

$$s_{oed} = m_v h \Delta u.$$

189

Table 9.2 *Increment in principal-stress ratio for footings*

h/b	Circular footing α	Strip footing α
0	1·00	1·00
0·25	0·67	0·74
0·50	0·50	0·53
1·0	0·38	0·37
2·0	0·30	0·26
4·0	0·28	0·20
10·0	0·26	0·14
∞	0·25	0

If lateral strain is taken into account $\Delta u = \mu p$ and the corrected settlement is

$$s = m_v h \mu p$$

that is

$$s = \mu s_{oed}.$$

For normally consolidated clays the factor μ is generally rather less than unity, while for overconsolidated clays the value may be down to 0·5 or less. In very sensitive clays the factor may be greater than unity.

9.4 PREDICTION OF SETTLEMENT

The term *settlement* is used to describe the vertical displacement of the base of a structure or of the surface of a road or embankment. The effects of settlement depend not only on its magnitude but also on its degree of uniformity and of the nature of the engineering work affected.

Settlement may be caused by:

(i) *Static loads,* such as those imposed by the weight of a structure or of an embankment;
(ii) *Moving loads,* such as are transmitted through a road or airfield pavement;

(iii) *Changes in moisture content,* which may arise from natural causes such as seasonal fluctuation in the water table or the abstraction of water by the roots of large trees. (Nearby excavation, pile-driving, pumping, or drainage may also have an important effect.)

(iv) *Undermining,* due to mining operations, tunnel construction, or underground erosion

Types of settlement
The settlement caused by compressive loading may be divided into two kinds:

(i) *Immediate settlement.* This is a combination of elastic compression and plastic deformation, without change in volume or water content, as shown in Fig. 9.9. This type of settlement develops as construction proceeds.

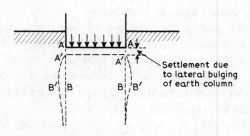

Fig. 9.9 Immediate settlement.

(ii) *Settlement due to consolidation.* This is the result of the decrease in the volume of the loaded soil caused by the gradual expulsion of water from the voids. In clay soils such settlement develops very slowly, but may attain considerable magnitude in course of time. Settlement is seldom uniform over the area occupied by a large building, because of non-uniformity of pressure distribution in the soil as well as variations in the compressibility at different parts of the area. Consolidation sometimes takes place in a compressible stratum situation at a considerable depth, although the soil immediately below the foundation is relatively firm. This may result in settlement of the foundation. In cohesionless soil settlement takes

191

place relatively quickly, and cannot so easily be separated from the 'immediate' settlement as defined above.

For an approximate estimate of the immediate settlement of structures, the elastic theory may be used. For consolidation settlement on clay soils the most reliable method is that based on compressive tests on undisturbed samples and the application of the theory of consolidation.

Loading tests can give useful information regarding immediate settlement, when used with caution and in the light of practical experience. For consolidation settlements, however, comparison of actual settlements with those predicted from loading tests show that this method is unreliable.

Immediate settlement
The theoretical settlement in an elastic medium can be developed from Boussinesq's equations for the displacement of a point load acting on the surface of a semi-infinite elastic medium. By integration or by the use of influence factors, the theoretical settlements caused by uniformly loaded area can be estimated.

Soils do not strictly obey the laws of elasticity; also they are not homogeneous. Elastic theory can be used for soils to predict with a fair degree of accuracy the pressures induced by loading, but it is not so reliable for the estimation of deformations. For cohesive soil, however, under suitable conditions, a rough estimate of the immediate settlement can be made, based on elastic theory.

For cohesive soil a simple approximate relation can be established between the settlement per unit width s/B and the strain measured in a compression test. The settlement s in an elastic medium under a loaded area of width B is given by an expression of the form

$$s = I_s q B \frac{1 - \mu^2}{E}$$

where I_s is an influence factor depending upon the shape of the loaded area and the distribution of contact pressure. Values of these influence factors are given in treatises on theory of elasticity.

For a uniformly loaded rigid circular footing applied to the surface of an elastic medium, the influence factor I_s is $\pi/4$. Assuming Poisson's ratio $\nu_u = 0.5$, the equation for settlement becomes

$$\frac{s}{D} = \frac{\pi}{4} \frac{(1 - 0 \cdot 25)}{E} q = \frac{0 \cdot 6}{E} q$$

where D is the diameter of the footing. E is usually taken as the secant modulus obtained from an unconfined or triaxial compression test over a range of stress from zero to one-half of the ultimate stress.

Settlement due to consolidation

The magnitude and rate of settlement of clay soils caused by consolidation are usually estimated by applying the theory of one-dimensional consolidation. For this purpose it is necessary to carry out soil surveys and to obtain undisturbed samples at various depths and from various parts of the site. The relevant properties of the soil are then measured by the oedometer test.

Sometimes valuable information can be collected from the results of settlement observations on existing buildings near the site, and it is desirable that much more data of this kind should be recorded. Such information cannot, except perhaps for small unimportant structures, obviate the necessity for carrying out thorough site investigation and soil testing.

It must be noted that under the conditions of the oedometer test there is no immediate or undrained settlement, since the specimen is confined laterally and can undergo change in volume only if water flows out or in.

In the conventional method for prediction of settlement there are some inconsistencies. The stress distribution with depth and the immediate settlement are calculated by three-dimensional theory, but in estimating the consolidation settlement a condition of unidirectional drainage is assumed

Magnitude of settlement

Consider a layer of saturated soil (of thickness h) beneath the foundation of a structure. We shall assume that this layer has been fully consolidated by the overburden pressure p_1. During the erection of a structure the pressure increases to $p_1 + p$. If m_v is the mean coefficient of compressibility for the range of pressure, as determined by a consolidation test, the volumetric change is $m_v p$. Neglecting lateral strain the settlement s is the reduction in the thickness of the layer, $m_v ph$.

It is assumed that lateral expansion is prevented and therefore all the compression takes place in a vertical direction. It is also assumed that drainage takes place either in one or in both vertical directions, but that lateral drainage is negligible.

The distribution of vertical pressure with depth under the loaded area is computed by the methods of Chapter 6. The effective pressure at any depth is then the pressure due to the load plus the original overburden pressure. For depths below the water table the overburden pressure is reduced by hydrostatic uplift or pore-pressure.

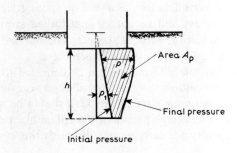

Fig. 9.10 Variation of vertical pressure under a loaded area.

To allow for the variation of pressure with depth a diagram is plotted showing the pressure distribution before and after application of the load (Fig. 9.10). If A_v is the area of the applied pressure diagram, then

$$\text{the average pressure} = \frac{A_p}{h}$$

and

$$\text{the settlement } s = m_v \frac{A_p}{h} h = m_v A_p.$$

To allow for the variation of m_v with the pressure at different depths the thickness h can be divided into thin layers, and the settlements of these computed and totalled. For any such element of thickness δh the mean pressure is taken as that at the depth of the centre of the element, and the coefficient m_v for this pressure is obtained from the m_v/p curve (Fig. 9.12). The settlement δs of the element is given by

$$\delta s = m_v p \delta h.$$

The compressible stratum should be divided into layers and the separate settlements calculated and summed.

Influence of size and depth of bearing area

On perfectly elastic homogeneous material the settlements under rectangular areas uniformly loaded at the same intensity are proportional to the width of the loaded area. Similarly, for circular areas the settlements are proportional to the diameter. Experiments have shown that the above-mentioned relation holds good approximately for cohesive soils, but for sandy soils the settlement under a given pressure tends to be more nearly independent of the size of the loaded area.

For footings below ground-level the settlement decreases as the ratio depth/width increases, but the greater the cohesion of the soil the less is the effect of this ratio.

Terzaghi has shown that about 80 per cent of the total settlement is due to consolidation of the soil within the pressure bulb bounded by the line representing a vertical pressure of one-fifth of the applied load intensity. By similarity it is apparent that the settlement should be proportional to the width, but the settlement of a large area as estimated from tests on a small area may be considerably in error if the pressure bulb penetrates into strata with different properties. The cumulative effect of neighbouring loads is likely to cause settlements in excess of those predicted from loading tests.

Rate of settlement

Fig. 9.11 shows typical time/settlement curves for buildings. Curve (a) shows the type of settlement for a structure on sand. Settlement proceeds during the period of construction and then ceases. Curve (b) is typical for clay, where the settlement continues for a long period, but at a gradually decreasing rate. The shape is similar to the consolidation/time curve shown in Fig. 9.1. In curve (c) the settlement continues to increase at a fairly uniform rate, probably due to plastic flow of material from underneath the structure. Curve (d) shows a spontaneous settlement due to some change in the soil conditions, such as decrease in strength caused by excessive moisture or by the influence of some neighbouring constructional work.

The time t required to reach a certain percentage consolidation of a stratum of thickness h is given by the equation

$$T_v = \frac{c_v t}{d^2}$$

where T_v is the time factor corresponding to the degree of consolidation, c_v is the coefficient of consolidation, and d is the length of the drainage path ($d = h$ for drainage in one direction only and $h/2$ for two-way drainage). The value of c_v used should be the average over the range of pressures involved, corrected for secondary consolidation, if feasible. The distribution of pressure with depth is taken as approximating to one of the standard cases shown in Table 9.1. The appropriate time factor T_v is obtained from the table, and the time t is calculated from the above equation

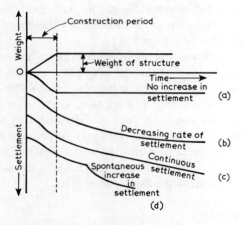

Fig. 9.11 Settlement curves.

The theory governing the rate at which three-dimensional consolidation takes place is complicated. As might be expected, the general effect is to reduce the time factor and so increase the rate of consolidation compared with the value calculated by the conventional unidirectional method.

Differential settlement
Although the calculation of total settlement is important, of even greater importance is the estimation of differential settlement. If a

building settles uniformly, each footing sinking by the same amount, and at the same rate, no structural cracking occurs. A large total settlement does not necessarily result in damage to the building.

If, however, two points in the structure settle by different amounts, the distortion or racking which occurs may be quite serious even if the amount of settlement is not large. This statement applies with the greatest force to statically indeterminate structures, where differential settlement can produce major structural damage.

There is no fixed relationship between total and differential settlement, even on a uniform soil. Buildings having a large total settlement are certainly more likely to have a large differential settlement, but this is not always so. Many factors contribute to differences in vertical movements – types of foundation, local variations in soil, drainage or lack of it and perhaps inadequate design.

Figures which are guides of limited but significant importance were derived by Macdonald and Skempton after studying the records of settlement of many buildings. Any angular distortion (difference in settlements divided by the distance between the points at which they are measured) above 1 : 300 causes cracking in finishes and claddings. Structural frames can accept 1 : 150 before structural damage results.

Effective load producing consolidation

In designing footings for a specified bearing pressure it is usual to estimate the worst possible condition of loading, e.g. a combination of dead load, live load placed to produce its maximum effect, and eccentricity due to wind pressure. It is unlikely that such maximum loads will act for any length of time, if ever, and where consolidation is taking place over a period of years we require the time-average of the load. There seems to be little data available for estimating this average, but a fair value should be obtainable by taking the dead load plus a percentage of the maximum live load, the percentage depending on the nature of the structure. The pressure in the compressible soil corresponding to this average load should be used in the settlement computations.

Hitherto we have assumed that the full load is instantaneously applied as in the oedometer test. Actually, the application of the load is spread over the period of construction. Terzaghi proved that the settlement caused by a pressure which increases at a uniform rate from 0 to p during a time t is the same as that occurring under a steady pressure acting for half that time. As the time occupied by

197

construction is usually small compared with that necessary for complete consolidation, it is usually justifiable to assume uniform pressure, taking the time origin half-way though the period of erection.

Settlement as deduced from penetration tests (see *In-situ testing*)
Settlement of the granular materials to which penetration tests apply, is seldom a serious problem in normal conditions. Its assessment is not as amenable to theoretical deduction as that of the settlement of cohesive materials. Theories of the settlement of granular soils have ranged from the correlation of SPT values with mathematical derivations of the settlement of a semi-infinite, isotropic, elastic medium, to comparison of the observed settlement of buildings with similar theoretical deductions.

If is agreed that the expression, generally known as the Terzaghi-Buissman equation is the most closely related to experience as present, although the subject has not yet reached a final conclusion. The basis of the expression is the value obtained from the static cone penetration test. In the expression, the following symbols are used:

k_1 overburden pressure at the mean depth of the stratum whose decrease in thickness is required

Δp increase in pressure at the same depth, due to the applied load of the weight of the building or other artificial construction

h thickness of the layer concerned

C_p Static cone penetration value (a pressure)

s settlement caused by the change in thickness of the stratum concerned.

$$s = \frac{hp_1}{C_p} \log \frac{p_1 + \Delta p}{p_1}.$$

This expression relates to one stratum, but the total settlement due to the reduction in thickness of several granular strata can be obtained by summation:

$$\Sigma s = \sum \frac{hp}{C_p} \log \frac{p_m + \Delta p}{p_m}.$$

If there are cohesive layers under the building, their settlements must be added to the non-cohesive settlement to obtain a measure of the settlement of the building

Settlement caused by repeated loads

Most soils exhibit some degree of elasticity under moderate stresses, in that most of the deformation disappears when the stress is removed. There is always, however, some residual strain, and considerable permanent deformation often results from the cumulative effect of repeated loading. A familiar example is the subgrade of a road subjected to repetitions of stress under the action of traffic. Every application of load causes deformation, and the recovery after the passage of any wheel is not quite complete. In compressible clay soils the consolidation properties contribute appreciably to the gradual settlement.

Settlements of this type, particularly if uneven, are liable to cause failure of pavements. For example, with concrete slabs the maximum deflections caused by passing wheel loads occur at the corners and at transverse joints. The soil at these places is therefore subjected to greater pressure and deformation than elsewhere, and progressive settlement may thus lead to the gradual withdrawal of the support of the soil, eventually causing cracking of the concrete slab.

Settlement caused by undermining

The effect of mining operations, such as the removal of rock salt or coal, is to produce a gradual subsidence of the ground overlying the working. The strata above the cavity tend to sag, causing not only large settlements but also the bending of the surface of the ground, accompanied by horizontal movements and forces both in tension and compression. The possible effects on structures situated in the area of subsidence may be summarized as follows:

(i) vertical settlement at the centre of the area;
(ii) vertical settlement combined with tilting;
(iii) non-uniform bearing pressures on foundations;
(iv) horizontal movements of the soil in regions of tension and compression.

As minerals are extracted over the width of an advancing face, as is the practice for coal, a wave of subsidence is apparent at the surface. The first portion of the waveform causes tension in the ground, the

199

middle portion causes compression, and the final portion again causes tension. The wave, as with waves in water, advances through the strata above the seam as the coal face advances. At the surface, the final position below the original position is the *subsidence* which is the most obvious result of mining operations. To the foundation engineer, however, the stresses developed by succeeding tensions and compressions are equally if not more important and should be remembered if subsidence is expected.

In coal mining, the amount of subsidence finally taking place depends on three factors:

(i) the depth of the seam below the surface;
(ii) the width of the advancing face;
(iii) the thickness of the seam extracted.

It is usual to use dimensionless factors in the estimation of subsidence. These factors are

$$\frac{\text{subsidence}}{\text{thickness of seam}} \quad \text{and} \quad \frac{\text{width of face}}{\text{depth of seam}}.$$

It is coincidental, but a useful mnemonic is that each of these simultaneously reaches a value of about one-half. The subsidence is approximately half the thickness of the seam if the width of extraction is half the depth of the seam below the surface.

The greatest proportional value of vertical subsidence occurs when the thickness of the seam is large and its depth below the surface is small in relation to the width of extraction. In such instances, the subsidence at the surface may be as much as seven-eighths of the seam thickness. This occurs when the width of extraction is approximately equal or greater than the depth of the seam.

These approximate figures refer to the situation when little or no attempt is made to fill the *goaf* or extracted volume. With the best methods of storing and packing available, the subsidence may, at considerable cost, be reduced to just under one-half of the maximum.

There are only two methods of diminishing damage due to settlement. The structure, on the surface, may be made so monolithic and strongly reinforced that it effectively resists the tiltings, tensions, compressions, and settlements which result from mining subsidence. On the other hand it may be made so flexible by specially designed joints that it offers no resistance to movement and so suffers no damage. Of the two methods the second is likely to be the more successful, and to cost less. It is very difficult and costly to make a structure sufficiently monolithic to resist all stresses imposed by subsidence. The 'rafts' on which faith is sometimes placed are often far too thin and weak to protect a building beyond a very small range of movement.

9.5 APPLICATION OF THE WORK OF CHAPTER 9

Where it is possible to calculate the effect of the properties of soil on the consolidation and settlement of a site under construction load, the student must be able to obtain the values of coefficient of compressibility and coefficient of consolidation from the results of an oedometer test. The former of these coefficients is the more important. From these values and the pressures applied at various levels below the surface (as found by the methods of Chapter 6) it is possible to progress from the values of the coefficients to the magnitude and rate of settlement in particular circumstances. The five problems give the methods to be adopted. Work through the problem again, and then alter them slightly in dimensions and loading and repeat the work until it becomes familiar.

Application 9A   Coefficient of compressibility
A sample of alluvial clay was subjected to a consolidation test, each increment of applied pressure being allowed to act until the sample attained its equilibrium thickness under that increment. The properties of the sample were:

> *Bulk density:* 1.9 Mg m^3 *Liquid limit:* 46.4%
> *Specific gravity of particles:* 2.69   *Plastic limit:* 21.6%
> *Initial thickness of sample:* 20.24 mm   *Moisture content at end of*
> *Final thickness of sample:* 18.82 mm   *test:* 0.27.

The pressures applied and the resulting changes in thickness measured at the equilibrium state at the end of each stage were:

$p(\text{kN m}^{-2})$ 0   12   24   56   109   219   438   875   0
(mm) -0.11 -0.10 -0.23 -0.36 -0.53 -0.74 -0.73 -1.47

Determine the value of the coefficient of compressibility.
The basic data collected from the sample are not all required in the calculation of the coefficient, but they give an opportunity of visualizing the type of clay by comparison of the Atterberg limits and

moisture content with known values.

Problems of this kind are always best solved in tabular form.

Table 9.3 *Determination of coefficient of compressibility*

Applied pressure p (kN/m^2) (a)	Change in thickness δh (mm) (b)	Change in pressure δp (kN/m^2) (c)	Thickness of sample h (mm) (d)	Ratio $\delta h/\delta p$ (m^3/kN) (e)	Coefficient of compressibility $m_v = \dfrac{1}{h}\dfrac{\delta h}{\delta p}$ (m^2/kN) (f)
0	–	–	20·15	–	–
12	−0·11	+12	20·04	$9·17 \times 10^{-6}$	458×10^{-6}
24	−0·10	+12	19·94	$8·33 \times 10^{-6}$	418×10^{-6}
56	−0·23	+32	19·71	$7·18 \times 10^{-6}$	364×10^{-6}
109	−0·36	+53	19·35	$6·79 \times 10^{-6}$	351×10^{-6}
219	−0·53	+110	18·82	$4·81 \times 10^{-6}$	256×10^{-6}
438	−0·74	+219	18·08	$3·38 \times 10^{-6}$	187×10^{-6}
875	−0·73	+437	17·35	$1·67 \times 10^{-6}$	96×10^{-6}
0	+1·47	−875	18·82	–	**

(i) The first three columns of Table 9.3 represent the measured values given in the statement of the problem, and obtained from the consolidation test.

(ii) The fourth column is started at the bottom with the final measured thickness of the sample after the load has been released to zero. By using the values of the change in thickness [column (b)], and the end thickness (18·82 mm), it is possible to work back to the original thickness before the test started. This value of 20·15 is not exactly the value (20·24 mm) measured, but is close enough to show that no serious arithmetical error has been made.

(iii) Column (e) is the quotient of columns (b) and (c).

(iv) The values of the *coefficient of compressibility* are obtained by dividing the figures of column (e) by those of column (d). In any of these calculations, the figures (b) and (h) must be altered to metres by multiplying by 10^{-3}.

The values of the *coefficient of compressibility* generally decreases as the pressure increases. To allow of a reasonable separation of the values of m_v for the lower pressures, a graph on logarithmic paper can be drawn for future reference. In this example, the final results are

shown on Fig. 9.12

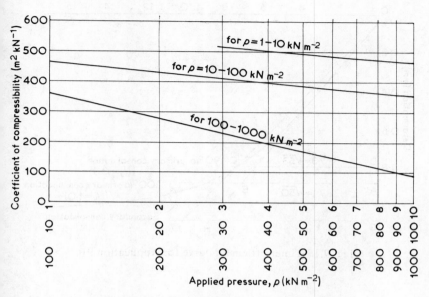

Fig. 9.12 $m_v p$ curves from Application 9A.

Application 9B   Calculations for oedometer test
From a sample of clay compressed in the consolidation test, the results were as follows: $d = 0.92$ mm

Time t from start of loading (s)	5	10	16	30	45	60	120	240
Deformation δh (mm)	0.31	0.40	0.51	0.69	0.75	0.84	0.90	1.00

The object is to calculate the coefficient of consolidation by the two alternative methods given in Section 9.2.

(i) Draw the experimental curve as in Fig. 9.13. In this curve the $\sqrt{t}$ values are:

2·2   3·2   4·0   5·5   6·7   7·7   10·9   15·5.

(ii) Draw a straight line through the upper part of the curve and

203

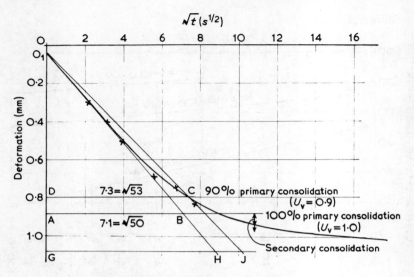

Fig. 9.13 Time/settlement curve for Application 9B.

project it to some point H well below the final level of consolidation. The upper end of this straight line does not pass through the origin. This is frequently found for, on the application of load, there is often an immediate settlement due to the compression of air voids. Deformation measurement can be taken from O_1 as the new origin of the curve.

(iii) Set off the length GJ so that $GJ = 1 \cdot 15\ GH$. Join JO_1.

(iv) Where JO_1 passes through the experimental curve is the point C which represents 90 per cent of primary consolidation.

(v) Draw the horizontal line CD, and locate A so that $O_1D/O_1A = 0 \cdot 9$. Where O_1H cuts the horizontal through A is the point B of Fig. 9.7 and 9.8, at the level of 100 per cent primary consolidation.

(vi) Measure $CD = 7 \cdot 3 = \sqrt{53}$
$$AB = 7 \cdot 1 = \sqrt{50}.$$

(vii)
$$c_v = \frac{\pi d^2}{4t_1} \quad \frac{\pi \times 9 \cdot 2^2}{4 \times 50} = 1 \cdot 33\ \text{mm}^2\ \text{s}^{-1}$$

$$c_v = \frac{T_v d^2}{t} \ (\text{for } U_v = 0 \cdot 9) = \frac{0 \cdot 848 d^2}{53} = 1 \cdot 35\ \text{mm}^2\ \text{s}^{-1}$$

Application 9C Coefficient of consolidation

The curve in Fig. 9.14 on semi-logarithmic paper shows the deformation/log-time relationship of a soil under a load, as the oedometer sample is compressed to equilibrium consolidation. Determine the coefficient of consolidation. Thickness of sample 19 mm. (Note that it is not necessary to measure changes in thickness in mm).

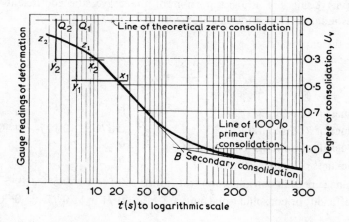

Fig. 9.14 Log time/deformation curve for Application 9C.

(i) The ordinates on the curve can be merely the dial readings of the gauges recording deformation.

(ii) At a point X on the curve near the top, draw a horizontal line. Mark on this horizontal line the time Y making it one-quarter of the time X.

(iii) Draw a vertical YZ and produce it to Q so that ZQ equals YZ.

(iv) Repeat this more than once towards the upper part of the curve to locate the theoretical line of zero consolidation. This is the line to which the curve would be asymptotic if it were possible to take accurate readings to very small values of time. It is obtained by joining Q_1, Q_2, Q_3, etc and producing. This construction is due to Casegrande.

205

(v) Draw a straight line through the middle portion of the curve. Also extend the straight or nearly straight portion of the curve representing secondary consolidation to meet the first line in *B*. In Figs. 9.7 and 9.8, the point *B* was defined with no regard to secondary consolidation. Here, at least some attention has been paid to this feature of consolidation curves. *B* defines the level of 100 per cent primary consolidation.

(vi) Divide the depth between the lines of theoretical zero and 100 per cent consolidation into 10 parts representing values of U_v.

(vii) Using the straight portion of the curve, the coefficient of consolidation can be found as follows:

Use the expression:

$$\frac{T_v d^2}{t} = c_v.$$

Values of U_v, say	0·3	0·5	0·7
Values of T_v from Table 9.2	0·071	0·197	0·403
Values of t from Fig. 9.14 using U_v	9	23	52
Thickness of sample 19 mm			
Drainage path in consolidation test (mm)	9·5	9·5	9·5
Coefficient of consolidation (mm^2 s^{-1})	0·71	0·77	0·70

Application 9D   Settlement under a circular footing
A surface bearing pressure of 240 kN m^{-2} is applied to a 2 m diameter circular pad footing. At a depth of 3 m there is a stratum of clay which extends to a depth of 3.6 m. The coefficients of compressibility of the clay are those of Fig. 9.12. What is the likely settlement of the pad footing due to the consolidation of this clay layer?

(i) Determine the pressure applied vertically at a depth of $(3 + 0·6/2)$ m below the surface load. This level represents the centre of the clay layer. The methods of Chapter 6 should be used. Using Table 6.2:

Ratio of diameter of pad to depth of stratum is $2·0/3·3 = 0·6$.

The influence factor for circular footings, related to the ratio 0·6, is 0·1213 (Table 6.2).

Pressure at 3·3 m depth is thus $240 \times 0·1213 = 29$ kN m^{-2}.

(ii) From Fig. 9.12, read off the coefficient of compressibility at this pressure: $m_v = 410 \times 10^{-6}$ m^2 kN^{-1}.

(iii) The consolidation of the clay layer is thus,

$$\delta h = m_v h \, \delta p$$

$$410 \times 10^{-6} \times 0.6 \times 29$$

$$0.007 \text{ m or 7 mm.}$$

The pressure of 29 kN m^{-2}, is registered as δp, a *difference* in pressure, as the object of the study is to find the difference in thickness of the stratum. This resulted from a difference of pressure of 29 kN m^{-2} from the original value which was caused by the weight of the soil itself. The other material below the footing was not mentioned in the statement of the problem; the inference is either that it is granular and shows little settlement, or the consolidation of these other layers is already known. The total consolidation of soil under a footing can always be found by adding the consolidations of the various layers.

If the pressure applied at the centre of the clay layer is greater or smaller than 29 kN m^{-2}, not only is δp changed in the expression for consolidation, but m_v is also different, and must be read off for whatever value of pressure is applied.

Application 9E   Settlement under a square footing

A square footing, of 3·5 m side is located at a depth of 2·5 m below the surface. A bed of clay extends from the surface to a depth of 10 m. Below this lies dense sand. The compressibility of the clay may be assumed to be that shown in Fig. 9.6.

An estimate is required of the total settlement of the centre of the footing.

(i) The depth of clay below the footing is 7·5 m. Divide this into a number of layers for convenience of calculation. A suitable number is five, each 1·5 m thick.

(ii) Determine the depths to the centres of the layers. These are shown in column (b) of Table 9.4. Make a sketch.

(iii) Use Table 6.4 (at the end of Chapter 6) to obtain the influence factors for the five depths.

$L = B = \frac{1}{2} \times 3.5$ m. Columns (d) and (e) show the influence factors for pressure at the corner of the half-width footing and the pressure at these levels. Since there are four half-width square footings making up the total fotting, the pressures developed must be multiplied by four to give the total pressure under the centre of the footing.

(iv) The values of the *coefficient of compressibility* are read off from Fig. 9.12. For other types of clay, graphs similar to Fig. 9.12 must be developed from the consolidation test, whereas Table 6.4 applies to all square or rectangular footings.

(v) The total settlement is summed from the individual settlements of the five layers.

It helps to sketch a cross-section of the footing and the foundation layers, writing in depths and values of m_v as found in the calculations. Draw the pressure/depth diagram A_p (see Fig. 9.10).

Table 9.4 *Settlement of a deep stratum*

Layer no. (a)	Depth z (m) (b)	Value of B/z (ratio) (c)	Influence factor (Table 6.4) (I.F.) (d)	δp (kN/m²) (e)	m_v (m²/kN) × 10⁻⁶ (f)	$m_v h \delta p$ (m) (h = 1·5 m) (g)
1	0·75	2·33	0·233	31·68	405	0·019
2	2·25	0·78	0·124	16·86	440	0·011
3	3·75	0·47	0·068	9·25	470	0·006
4	5·25	0·33	0·041	5·58	485	0·004
5	6·75	0·26	0·023	3·13	510	0·002
					Total settlement:	0·042 m

The values of δp are obtained from I.F. x 4 x 34. The value of I.F. x 34 represents the stress under the corner of the half-width footing. The total stress comes from four such corners. The values of m_v are read off from Fig. 9.6 which has been extrapolated to the lower values of δp to give the approximate values of m_v.

The settlement can also be obtained, with as close an accuracy as need to be looked for, by taking the area of the pressure diagram, represented by the sum of the values δp multiplied by 1·5 m. This is 99·75 kN m⁻¹. The mean of the values of δp is about 17 kN m⁻² and a glance at Fig. 9.6 shows that at this pressure the value of m_v is about 440 m² kN⁻¹. The total settlement is, therefore,

$$A_p m_v = 99\cdot75 \times 440 \text{ m} = 0\cdot044 \text{ m}$$

when a mean value of m_v is assumed. This is so close to the more elaborately calculated value that the extra effort is probably unjustifiable unless the conditions are complex, with different materials contributing to the settlement. Here the material is uniform.

Application 9F   Rate of settlement
Find the rate of settlement of the 7·5 m stratum in Application 9E and the time taken to reach 90 per cent primary settlement. The value of c_v has been measured as 35 × 10⁻³ mm² s⁻¹.

(i) Column (e) of Table 9.3 shows the pressure decreasing from a maximum to nearly zero. The nearest type of standard distribution is Type 2 of Table 9.2. This is triangular diminishing in the direction of drainage. The drainage is in one direction and the drainage path 7·5 m.

(ii) From Table 9.2, Type 2 (note permeable is shown at top instead of at bottom) the value of T_v for 90 per cent primary consolidation is 0·940.

(iii) The value of the time taken to consolidate 7·5 m to 90 per cent primary consolidation is, therefore,

$$t = \frac{T_v d^2}{c_v} = \frac{0\cdot94 \times (7\cdot5 \times 10^3)^2}{35 \times 10^{-3}} \text{ seconds}$$

$$= 48 \text{ years.}$$

CHAPTER 10

Foundations

When a structure is constructed in or on the soil, it applies stresses in the soil in excess of what existed originally. Assuming that the original condition of the soil was stable, excess stress may cause deformations in the soil large enough to be deleterious to the structure applying the load. It is the object of the designer of foundations to ensure that such deformations do not take place.

The structure may be a building, a bridge, a tunnel or may be constructed of the soil itself – an embankment or, negatively, a cutting. Although the effect of the excess load applied by an embankment or the release of load caused by the construction of a cutting are foundation problems, the techniques of dealing with structures composed of soil are dealt with in another chapter. In this chapter, emphasis is on the foundations of artificially constructed superstructures. The chapter deals only with the soil; the pads, footings or foundations constructed as part of the building are a matter for structural analysis. The geometry and arrangement of these foundations which apply the load are, however, of importance and have significance in design. They may be divided into three classes:

(i) *Strip Footing* as for the base of a wall. From this type of fotting the load from the wall is spread through the soil at right angles to the line of the wall.

(ii) *Pad Footing* such as that for a pier or a column, where the effect of the load applied spreads in all directions through the soil.

(iii) *Raft* where a number of walls or columns are supported on a structurally stiff system of continuous slabs and/or walls. A particular category of this type is the *buoyant* or *cellular foundation* in which

210

the excavation is carried to some depth so that the weight of the soil extracted is wholly or partially equal to the weight of the building, and the excess pressure applied to the soil is thus small.

10.1 FAILURE OF FOUNDATIONS

When a mass of soil is loaded by the weight of a superimposed structure through a specially constructed foundation to that structure, the soil should not display any stress or deformation which can be classified as 'failure'. 'Failure' can be described in various ways. At one end of the scale, the slightest deformation (caused by movement of the foundation) of a building carrying delicate decorative finishes could result in unsightly cracking which, although not endangering the stability of the building, could be classified as failure, requiring remedial measures. At the other extreme, a much greater movement of the foundations of a large warehouse or hangar could take place without there being any visible distress in the structure. Indeed, very large settlements can occur without showing the signs of 'failure', provided all parts of the structure move at the same rate and through the same amount. This has occasionally occurred in areas of mining subsidence. School buildings, specially constructed to respond to large movements without distress, have settled, relative to their original position by amounts of 500 mm or more, without either teachers or pupils being aware of the movement.

Failure of the foundation material – the soil supporting the strip, pad or raft – can be classified into two chief categories.

Shear failure

This type of failure is similar to that experience in an embankment, and studied in another chapter. Shearing of the soil occurs along curved failure surfaces. Such failures are common in embankments or cuttings, but unusual under foundations. The possible disastrous effects on a large building are so well understood that shear failures are avoided by wise designers of foundations.

However, in order to avoid shear failure, it is important for the designer to know at what pressure a shear failure would take place. This pressure is a basic figure of some importance. It is called the *ultimate bearing capacity* of the soil. This is not a fundamental property of that soil, as many architects and engineers lightly assume. Its value depends not only on the type of soil and on the pressure applied, but also on the type and size of the foundation applying the load, and on the depth below the surface at which it acts. Clearly, the determination of the *ultimate bearing capacity* of the soil in particular circumstances, and the avoidance of any working load even approaching this figure, is of paramount importance.

Slow-deformation failure
This is the second important type of failure of the soil under a foundation. Here the failure is not spectacular, and even may show but little movement which is openly visible at first glance. The deformation, although slow, is continuous and cumulative. It is found when the soil is of the cohesive type where deformation occurs as moisture is extracted from the soil. This extraction may occur though evaporation in dry weather, through extraction by the roots of tress, by newly installed drainage or pumping installations, and particularly by the slow expulsion of water under the load of a structure. Of these four typical causes of slow deformation, the first three can be avoided and should not be allowed to occur. The fourth, consolidation under load, cannot be avoided but must be managed so that it does not constitute a 'failure'.

Differential settlement
Settlement, the slow-deformation distortion which cannot be avoided, need not cause failure or damage if properly controlled. If a building settles uniformly, each footing sinking by the same amount, and at the same rate, no structural damage should occur. A large total settlement does not necessarily result in damage.

However, the calculation of total settlement is of importance and is the first step to the control required. Of greater importance, however, is the *differential settlement*. If two points in a structure settle by different amounts, the distortion or racking which occurs may be serious even if the amount of total settlement is not large. The ideal situation mentioned above, when all parts of the structure settle equally, seldom occurs. It can be arranged by the use of expensive *raft* or *cellular* foundations, but the usual condition is that differential settlement will occur and must be controlled so that distortion does not constitute failure for the type of building in question. The control of differential settlement is of particular importance for statically indeterminate structures, where major structural damage could occur.
Unfortunately, there is no fixed relationship between total

settlement and differential settlement, even on a uniform soil (and foundation soil is hardly ever uniform). Buildings having a large total settlement are certainly more likely to have a large differential settlement, but this is not always so. Many factors contribute to differences in vertical movement: types of soil, local variations, drainage or lack of it, and perhaps inadequate design.

Figures which are at least guides to safe design have been extracted empirically from the records of buildings which have shown and which have not shown structural damage. They have been reported by Macdonald and Skempton. The guide lines they suggest are:

Determine the angular distortion between two parts of a structure by comparing settlements. The angular distortion is the difference in total settlement of the two positions divided by the distance between them

$$\text{The ratio } \frac{\text{differential settlement}}{\text{distance}} \text{ should not exceed}$$

1 in 300   if cracking of finishes and claddings is to be avoided

1 in 150   for more causes damage to main structural members.

Conditions for avoiding failure

To summarize, there are two checks to be made on any proposed foundation design:

(i) Has the *ultimate bearing capacity* been divided by an adequate factor of safety to obtain the working pressure the foundations may apply with safety and the avoidance of shear failure?

(ii) Are the *ultimate settlements* of individual units of the foundation of such values that the *differential settlements* between adjacent parts of the structure lie within acceptable limits?

In answering these questions, particular attention must be paid to cohesive soils. Dense soils having a considerable angle of shearing resistance are unlikely to cause much difficulty; their ultimate bearing capacity is high, and settlements take place while the load is being applied, showing no slow deformation.

10.2 SHEAR FAILURE AND BEARING CAPACITY

The attempts which have been made to determine the load at which a sudden shear failure will take place under a foundation (*ultimate bearing capacity*) can be classified into four groups, the pressure at which failure occurs being estimated by one or more of these methods:

(i) a study of the active and passive pressures occurring under the footing;
(ii) an analysis of plastic shear failure;
(iii) a semi-graphical method, assuming a probable shape for the slip surface (see Chapter on stability of slopes for the fundamental conceptions);
(iv) direct measurement by loading tests.

In all these, the shearing strength of the soil under the footing is of first importance. The type of material most susceptible to shear failure is one in which the immediate shearing resistance does not increase with increasing loads (clays and silts). Materials with large angles of shearing resistance are not so likely to fail by shear, since shearing resistance increases with the increase in the superimposed load, and depth.

Methods based on active and passive earth pressures
In Rankine's well known theory for the minimum depth of foundations in cohesionless soil, the vertical downward pressure of the footing is considered as a maximum principal stress, and the lateral or minimum principal stress is the corresponding active earth pressure. This lateral stress is, for particles just beyond the edge of the footing, considered as a maximum principal stress, which in turn brings into play a vertical minimum principal stress. Rankine's methods of evaluating the principal stress causing shear failure in a cohesionless soil are of classical interest, but are open to a number of serious objections. They always give results lower than those found from tests. An abrupt change in stress conditions is implied below the edge of the footing, and this is contrary to the facts. The bearing pressures thus calculated are independent of the size and the shape of the footing, a result which again conflicts with actual conditions.

The type of failure occurring in plastic material is better

exemplified by movements such as those indicated in Fig. 10.1. The shaded wedge may be imagined to be pushed down intact, producing lateral thrusts which overcome the passive resistance of the soil lying to each side of the footing. The depth of the wedge decreases as the roughness of the base of the footing increases.

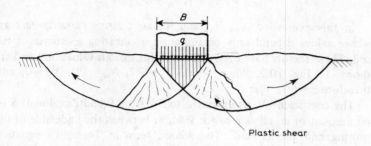

Fig. 10.1 Plastic failure

Plastic failure and bearing capacity factors
From the theory of plasticity Prandtl developed expressions for the ultimate bearing pressures on soils under a strip footing, the load being applied at the surface of the ground. The curved part of the slip surface is part of a logarithmic spiral. For the stability of foundations on saturated clay, immediately after completion of construction, it is a common practice to assume the undrained condition, that is, $\phi = 0$. For cohesive soil, on the $\phi = 0$ assumption, the spiral becomes a circular arc and Prandtl's analysis leads to the equation:

$$q = (\pi + 2)c = 5 \cdot 14c.$$

The ultimate bearing capacity on cohesive soil is therefore independent of the width of the footing, but on frictional soils Prandtl's theory indicates that it increases with the width.

Working on similar lines, Hencky analysed the problem of a uniformly loaded circular area on the surface, and found that

$$q = 5 \cdot 64c.$$

Both these derivations were made for loading at the surface; they are not applicable when the footing applies its load at a depth much below surface level.

Terzaghi investigated the problem using somewhar similar assumptions as to the mechanism of failure, and allowing for friction and cohesion between the base of the footing and the soil. He gives the following expression for the bearing pressure under a shallow strip footing which will cause general shear failure:

$$q = cN_c + \gamma z N_q + 0 \cdot 5 \gamma B N_\gamma$$

In this expression N_c, N_q, and N_γ are *bearing capacity factors*, whose values depend only on the angle of shearing resistance of the soil. These factors have been calculated for various values of ϕ and are shown in Fig. 10.2. When $\phi = 0$, $N_c = 5 \cdot 7$, $N_q = 1 \cdot 0$, $N_y = 0$, and therefore $q = 5 \cdot 7 c + \gamma z$.

The coefficient N_c is increased to $5 \cdot 7$ from Prandtl's value of $5 \cdot 14$ on account of an allowance for friction between the underside of the footing and the ground. The second term in Terzaghi's equation represents the increase of bearing capacity due to the overburden, and it will be seen that for frictional soils N_q increases rapidly with ϕ. The third term, which applies only to frictional soils, indicates the effect of the breadth of the footing on the bearing capacity.

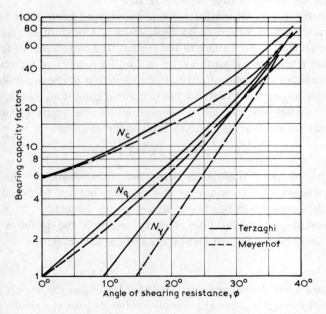

Fig. 10.2 Bearing capacity factors: strip load close to surface

Meyerhof has developed and expanded Terzaghi's method, but the main differences lie only in the exact values of the bearing capacity factors. Fig. 10.2 shows the comparison between Terzahi and Meyerhof and this diagram can be used to determine the bearing capacity factors for any angle of ϕ. When the material is a purely cohesive clay, and $\phi = 0$, the two sets of values agree. It is with this type of clay that the most severe problems arise.

Slip-surface methods

The slip-surface methods of the chapter on stability of slopes could be adapted to the estimation of the bearing capacities of footings.

The best-known method is that known as the 'circular arc' method, originally proposed by Fellenius for a strip load applied at the ground surface, and subsequently modified for footings founded below ground level. Failure is assumed to take place by the heaving up of a mass of soil on one side only. Model tests and observations of failures in practice show that because of slight eccentricity of loading or lack of homogeneity of the soil the failure is often one-sided instead of symmetrical.

This method and similar slip-surface methods are now merely of academic and historical interest for foundations, as bearing capacity is now invariably estimated by the use of *bearing capacity factors*.

Foundations on granular soils

The bearing capacity of foundations on granular soils can be calculated from *bearing capacity factors* (Figs. 10.2 and 10.4 for a deep foundation). The result obtained (divided by a factor of safety) cannot be accepted without question as an allowable bearing pressure; it has been found by experience that the possible settlement of granular material under a load is often more significant as a controlling factor. If the value of allowable bearing capacity (ABP) obtained by the use of bearing capacity factors were used in design, the settlement would often be unacceptable, especially if the soil were saturated. This section shows how foundations may be designed by considering possible settlement. The results should then be compared with allowable bearing pressure from bearing capacity factors, and the lesser figure assumed for design.

There are at least four important influences which control the value of the ABP. These are summarized in Table 10.1. They give factors by which a basic ABP should be multiplied in order to

Table 10.1 *Adjustments to allowable bearing pressure on dry and moist sand*

	The effect	Action required for least settlement	Factors to be applied to previously calculated ABP
Width of footing	The narrower the footing, the less the settlement.	Use as narrow a footing as possible and consistent with good design.	Use Fig. 10.15 to correct first width selected.
Depth of water table	The deeper the water table below footing, the less the settlement.	Maintain water table as deep as possible.	WT below twice breadth of footing: no adjustment. WT at base of footing – saturated soil – x ABP by 0·5.
Saturated material	The finer the saturated material, the greater the settlement. The coarser the material the less the settlement.	Prevent material from being saturated by adequate drainage before construction.	For fine sand x ABP by 0·5. For gravel and sand mixtures x ABP by 2·0.
Depth of base below the surface.	The deeper the base of the footing below the surface, the less the settlement.	Found footing as deeply as economically possible.	If ratio of depth of footing to its width is unity x ABP by 0·67. As footing is founded nearer the surface, the factor is reduced as far as 0·5.

determine the design figure. The basic ABP is, therefore, of first importance.

The bearing capacity of dry or only moist sand (not saturated) is dependent chiefly on the density of packing. Naturally, a loose packing will result in a greater settlement. If the packing is below the *critical density* (see Index), the settlement can be large and rapid. The density of packing is measured by the *standard penetration test* (SPT). Here, it is sufficient to say that the larger the standard penetration value, the greater the resistance provided by the soil to penetration. In Table 13.1, the SPT value is correlated with three properties of granular soils. It is important to note that the *allowable bearing pressure* listed is that which produces a settlement of about 25 mm in dry or moist sand. For other conditions, factors must be applied to the ABP in Table 13.1.

The following descriptions, and the summary in Table 10.1 show how unwise it is to accept the basic ABP without considering other physical conditions on the site. The use of Table 13.1, Table 10.1 and Fig. 10.15 must be practised if the student is to become confident in the design of foundations on granular soil. There is no problem here of long-continuing settlement. The settlement is immediate, but can be substantial and even unacceptable unless precautions are taken.

Four of the factors which might be considered in order to assess their influence on the modification of the basic ABP are:

(i) *Width of footing.* As the footing on granular soil becomes wider, the bulb of pressure it creates encompasses a greater mass of soil (see Fig. 6.5) and thus a greater settlement is produced. To maintain the 25 mm settlement assumed to be acceptable, it is thus necessary, as the footing widens, to reduce the bearing pressure Fig. 10.15 (in the Applications section at the end of the chapter) shows how the allowable bearing pressure for a given SPT decreased with an increase in width of footing.

(ii) *Depth of water table.* The presence of a water table or phreatic surface within the granular material indicates a reduction in the effective stress between particles and, if there is a flow of water through the soil, may also produce a seepage pressure. So long as the depth of the water table below the base of the footing is about twice the width of the footing, no adjustment of the figures given in Table 13.1 is necessary. If, however, the water table rises to approach the footing, the bearing capacity of sand is much decreased.

(iii) *Saturated material.* Here we are concerned not so much with

219

the saturation, which is considered in the paragraph above, but with the type of material. If the material is dry, the difference between its bearing capacity and that of the dry or moist sand of Table 13.1 is not of such importance. The saturation of a fine sand, however, can considerably reduce its already low bearing capacity. For inundated gravels and coarser sands, the weakening effect is not so apparent, and such coarser granular materials can be stronger in resistance than the sand for which the columns in Table 13.1 are prepared. The factor by which the ABP must be multiplied is then greater rather than less than unity.

(iv) *Depth below surface.* If the base of a footing is well below the surface, the surcharge surrounding the foundation assists the granular soil in building up a greater resistance to penetration and settlement. If, on the other hand, the footing is founded near the surface, such assistance is not provided by the overburden, and the latent power of resistance is much reduced.

All these four factors are well recognized as controlling and modifying influences on the basic allowable bearing pressure. The procedure in designing a footing to carry a column load, for example, is to use the SPT value to find an appropriate ABP or range of ABP from Table 13.1. The multiplying factors of Table 10.1 are then applied.

Loading tests
It has long been the practice of engineers to carry out loading tests on the site in order to determine the bearing capacity of the soil or to check whether the bearing pressure adopted for a design is safe. Load is applied to a block of concrete or a steel bearing plate by means of dead weights or by a hydraulic jacking system.

The estimation of bearing pressures from loading tests is open to some serious objections. As has been shown in the diagram of pressure bulbs, in Fig. 6.5, a loading test may fail to disclose a soft stratum several feet below the bottom of the test pit. In addition, the ultimate bearing capacity is not always directly proportional to the area. The use of loading tests for the estimation of settlement is often even more misleading.

Deep foundations
The bearing capacity factors as devised by Terzaghi refer only to shallow foundations. By 'shallow' is meant depths which are less than

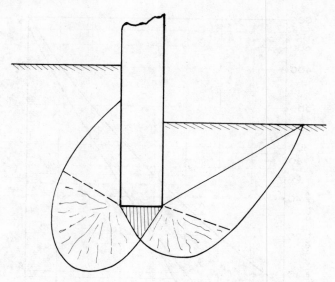

Fig. 10.3 Plastic failure: deep load

the breadth of the strip footing to which the equation refers. The factor N_q controls the effect of the overburden pressure. The overburden pressure allows frictional resistance to develop outside the immediate base of the footing and so gives an extra contribution to the stability of the foundation. The resistance increases as ϕ increases. When the depth at which the footing is placed is greater than its breadth and the overburden pressure, at the level of the base of the foundation, much increased, different patterns of failure ensue (Fig. 10.3), and higher values of the bearing capacity factors are used (Fig. 10.4). When the foundation penetrates to a considerable depth below the surface, the weight of the overburden around it alters the pattern of the zones of plastic shear failure. Meyerhof, again, has studied this condition, and Fig. 10.3 gives his conception of the pattern of failure. Naturally, from the condition shown in Fig. 10.1 there can be many intermediate patterns. The effect of the overburden causes an increase in the bearing capacity factors and thus in the net ultimate bearing capacity. Fig. 10.4 shows Meyerhof's estimate of the bearing capacity factors for net ultimate bearing capacity for a deep strip foundation.

Skempton has given a comphrehensive review of the various theories of bearing capacity, together with experimental data from

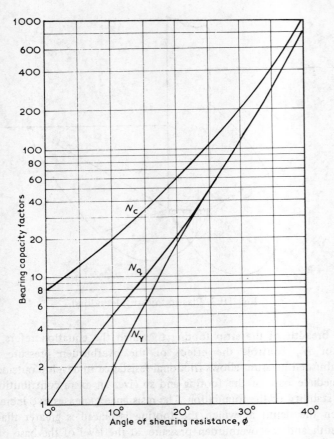

Fig. 10.4 Bearing capacity factors: deep strip load

both laboratory and full-scale observations. He forms the general conclusion that, for cohesive soil, the factor N_c increases with the depth excavated up to a maximum of about 7·5 for depths exceeding $2\frac{1}{2}$ times the width of the footing. His suggested curves are plotted on Fig. 10.5. It should be noted that the curve for strip footings starts at Prandtl's value of 5·14 for surface loading.

Whether the footing is shallow or deep, a weight of soil must be removed before the footing can be built. This soil had originally applied a pressure equal in its weight to the level of the base of the footing. To get the safe *net* bearing pressure, the value of q, as

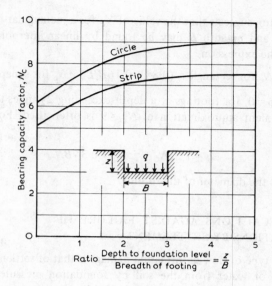

Fig. 10.5 Bearing capacity factors for cohesive soils

obtained from Terzaghi's expression, is divided by the factor of safety, F. A further pressure may then be added, equal to the weight of the column of soil which was excavated to allow the foundation to be inserted, and the total safe bearing capacity becomes

$$\frac{q}{F} + \gamma z.$$

Loading over an area

The basic equations for bearing capacity are based on the consideration of a strip footing – the two-dimensional case. If the footing is not a strip, but is square, rectangular, or circular, the support given is by a three-dimensional bulb of pressure, effective at the ends of the rectangle, or all round the circumference of the circle. The bearing capacity of such footings may be expected to be greater than that for a strip load. The various proposals for extrapolation are as follows.

For square footings Terzaghi suggests

$$q = 1 \cdot 3 c N_c + \gamma z N_q + 0 \cdot 4 \gamma B N_\gamma.$$

Skempton suggests that the value of N_c for a rectangular footing of length L, and breadth B may be found by linear interpolations, as given by the expression

$$N_c \text{ for rectangle} = (0 \cdot 84 + 0 \cdot 16 B/L) \times N_c \text{ for square.}$$

When $\phi = 0$, for footings at a depth exceeding $2\frac{1}{2}$ times the width, the convenient approximation of $N_c = 9$ is often used. For circular footings

$$q = 1 \cdot 3 c N_c + \gamma z N_q + 0 \cdot 3 \gamma B N_\gamma$$

where B is the diameter of the footing.

10.3 PRECAUTIONS AGAINST FAILURE OF FOUNDATIONS BY SETTLEMENT

The chief type of slow-deformation damage is that of settlement – the expulsion of water from the soil by foundation pressure, and the resulting decrease in void ratio. This is fully discussed in Chapter 9 which should be read in conjunction with this section.

As has been shown in Chapter 9, it is possible to obtain some estimate of the probable settlement of a foundation under a given load. It must be emphasized, however, that even under the most favourable conditions such estimates can be only approximate. The problem then remains to design the structure and its foundations to be safe against damage from settlement.

A common fallacy is that settlement can be prevented by spreading the foundation to produce a sufficiently low bearing pressure. While this procedure may help when the base of the structure rests directly on compressible soil, it is of little value when consolidation takes place in a deep-seated stratum. When the depth to the compressible stratum is several times the width of the uniformly loaded area, the vertical pressure can be calculated without appreciable error on the assumption that the load is concentrated at the centre of the area. At such depths, therefore, it is the total load, and not the bearing pressure under the footing, which is the dominant factor in producing settlement.

Preventive measures may be classified as follows:

   (i) prevention of excessive settlement;
  (ii) acceleration of settlement by sand drains;

(iii) design of structures to resist secondary stresses induced by settlement.

Prevention of excessive settlement of foundations

It has been shown that where settlement from the consolidation of relatively shallow strata is expected, it can be reduced by increasing the area of the foundation, but obviously there are limits to this procedure. A more economical as well as a more satisfactory solution may be to carry the footings down to firmer soil at a lower level. The use of pile foundations for this purpose is considered elsewhere, as is also the method of buoyant foundations, where the weight of the structure is partially compensated for by the excavation of basement space.

As regards soil properties the two conditions to be satisfied in the design of foundations are (i) the bearing pressure must not exceed the ultimate bearing capacity divided by a suitable load factor and (ii) the settlement must not exceed a specified value which depends upon the type, design, and purpose of the structure.

From the expression for the settlement in an elastic medium it is seen that the allowable bearing pressure for a specified settlement is inversely proportional to the breadth B of the footing. It can be shown that this applies also to the consolidation settlement when the structure is founded directly on compressible soil. Fig. 10.6 shows diagrammatically the variation of allowable pressure with breadth of footing, taking into account safety both against shear failure and against excessive settlement.

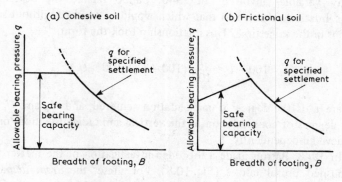

Fig. 10.6 Allowable bearing pressure

Sand drains

When large structures, sensitive to movement which accompanies settlement, are constructed over a clay stratum, cracking and damage to pipelines may continue over a long period. Earth dams are among the larger structures particularly affected by such movement and have failed because of the failure of underlying clays. Dams and fills apply considerable load to the soil, and even deep-lying clay strata are consolidated by such loading. Many years may elapse before settlement ceases and the maximum value of the movement is achieved.

For large structures, whose effectiveness depends on their stability, it would be of great advantage if maximum settlement could be reached earlier. The provision of shorter drainage paths for the escape of water to strata of higher permeability is one way of achieving this aim. The device by which more rapid draining might be effected was called by Terzaghi the *vertical filter well* and by Barron the *drain well*. These wells are now more usually referred to as *sand drains*.

Vertical bored wells, of the order of 300 to 600 mm in diameter, are filled with permeable sand and serve as supplementary outlets for the excess water. The spacing of the wells on a grid (Fig. 10.7) reduces the length of the horizontal drainage path to a value equivalent to half the distance between the drains. The result hoped for by this technique is the accomplishment of up to 80 per cent of the final settlement during the construction stage of the work. It must be admitted that this is not always achieved.

The mathematical argument relating the spacing and pattern of the drains to the resulting rate of consolidation is complex because of the many variables involved. In 1940, Carillo related the degree of consolidation achieved to that which would be reached without sand drains in the same time. This relationship took the form

$$100 - U = \frac{1}{100} (100 - U_v)(100 - U_r)$$

Where U is the degree of consolidation achieved, and U_v and U_r are the degrees of consolidation in the vertical and radial directions only, achieved independently.

In 1948, Barron made a thorough investigation of the matter and published useful curves (Fig. 10.8). For these, the *drain diameter ratio, n,* is defined as the ratio of the diameter of the influence area to

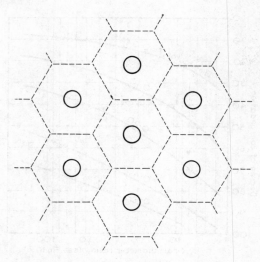

Fig. 10.7 Pattern of sand drains

the diameter of the drain. In Fig. 10.7 the influence area would be taken as a circle inscribed in one of the hexagons.

The only viable method of determining the rate of consolidation produced by sand drains, at present, is by the use of Carillo's equation and Barron's curves. An able summary of the state of progress was published by Christie in 1959, together with a long bibliography and a useful list of sites where, up to then, installations had been made.

All experiences have not been acceptable. If a very compressible material is concerned – for example, peat – the sand drains, being much more rigid than their surroundings, may take on the function of piles and restrict rather than increase consolidation. On the other hand, there has been a greater volume of reports of successes achieved. From may sites all over the world there have been records of a marked increase in the rate of consolidation witnessed. Unfortunately, few sites are provided with control areas where the rate of settlement without drains can be measured. Where this has been done, the results have sometimes not shown an outstanding acceleration in settlement due to the drains. Numerical examples on the problem of settlement with and without sand drains are fully developed in the Authors' *Problems in Engineering Soils*.

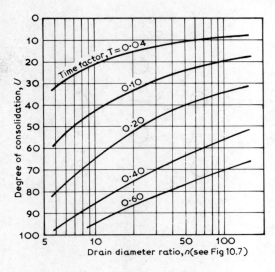

Fig. 10.8   Design chart for spacing of sand drains

Design of structures to resist settlement

The design of structures to resist settlement is outside the scope of this book, but a brief review of the principles involved will not be out of place here.

As a certain amount of settlement is often inevitable, provision for this must be made in design. The effects of differential settlement depend to a great extent on the rigidity of the structure. For example, a girder bridge of several freely supported spans is statically determinate, and unequal settlement of the piers, unless excessive, will not affect the internal stresses. In a continuous girder bridge even a small amount of differential settlement of the piers will alter the reactions, and consequently the shearing forces and bending moments on the girders. The effect of unequal settlement is even more pronounced in the portal frame type of reinforced concrete structure. In some instances a building is constructed upon a raft foundation so that the structure as a whole forms a rigid framework. Here considerable secondary stresses may be set up by uneven settlement, but the frame, acting as a very deep girder, can be more readily designed to resist such stresses.

In applying these principles to the design of large buildings, the structure may be supported on a number of independent footings or, alternatively, on a continuous foundation. In the former case the building should be of relatively flexible construction so that the reactions are determinate, and unequal differential settlement does not affect the stability. Even so, such differential settlement should be limited, as cracks may occur in plaster, floors may go out of level and other faults may develop, even though the stability is not impaired.

For large buildings a uniform spacing of uniformly loaded columns does not give uniform settlement. Such a loading results in a dish-shaped settlement. A non-uniform distribution of columns arranged so that those more heavily loaded lie towards or on the perimeter of the building results in decreased differential settlement.

The foundation for a circular water tank on clay requires quite a different treatment. The bottom of such a tank should be as thin and flexible as possible in order to avoid the high edge stresses caused by a rigid floor, for differential settlement between the rim of the tank and the central portion would cause cracking. Through a thin flexible floor, however, the uniform weight of the water tends to be transferred uniformly to the foundation.

Buildings of a rigid type should be supported on a reinforced concrete raft of adequate strength to resist the bending moments encountered; alternatively the footings of columns should be connected up by a series of continuous girders, so that the structure settles as a whole. Unequal settlement will then affect the distribution of the supporting reactions, and the structure must be designed to be safe under the worst possible conditions.

Locating causes of structural damage due to settlement

In mining areas, any structural damage is likely to be blamed on mining subsidence. But even in mining areas there are often other causes, and a detailed study of all circumstances should be made. For example, recent changes in drainage conditions (either by construction of a new drainage system or by excavation in or pumping from granular material) can cause serious damage. If subgrades are opened to long periods of dry or wet weather before foundations are laid, the attainment of the natural equilibrium moisture content, several years after construction, can result in sudden and apparently inexplicable cracking of buildings. A study of weather records for the time of construction is of value in such instances. The restraining of a statically determinate frame (e.g. a three-pinned portal) by later additions to the structure have resulted in damage in areas where subsidence (which would have had no effect on the statically determinate frame) did occur. If the depth of compressible material below individual footings varies considerably, there is a greater danger of differential settlement. The heavy extraction of moisture from clay soils by tree foliage, especially in dry summers, has often been the cause of structural damage. The establishment of the cause of structural damage is not simple, and the apparently obvious solution is not always the true one.

10.4 DESIGN OF FOUNDATIONS FOR STABILITY

It has been shown that failure of a foundation may take the form either of collapse by general shear or of a local shear failure. Also, under certain conditions differential settlement is liable to cause trouble even though no actual failure of the foundation material takes place.

The first problem, then, is to carry out a site investigation and soil tests to determine the nature and properties of the soil affected by the project. For settlement analysis the compressibility and consolidation properties are required, while for estimating the ultimate bearing capacity we require either loading tests or, preferably, the measurement of cohesion and shearing resistance by means of direct or indirect shear tests. Various methods of estimating the ultimate bearing capacity have been discussed in previous articles, and it now remains to formulate the procedure for design of foundations using the data available from tests.

The points to be considered are:

  (i) depth of footing;
  (ii) load factor against failure;
  (iii) area and shape of footing.

The design of the footing itself is outside the scope of this book.

Depth of foundations
The shrinkage of soil, particularly clay, is a frequent cause of trouble with shallow foundations. To avoid this, foundations of buildings

should be carried down to a minimum depth below which there is little likelihood of variation of moisture content. In temperate regions, such seasonal fluctuations rarely extend to a depth greater than 2 m, and the variation below 1 m is usually small. Other factors which may cause alteration of moisture content include fast-growing trees, nearby excavations, and underground workings. The possibility of the drying out of the foundation soil by boilers situated in the basement of a building should not be overlooked.

In cold climates the minimum depth is sometimes governed by the depth to which frost is likely to penetrate, especially where the soil is fine sand or silt. Artificial freezing may occur under buildings containing large refrigeration plant. When the bearing capacity of the soil increases with depth as in frictional soils, or where a firmer stratum underlies soft material, the depth to which the foundation should be taken may be governed by the location of a stratum of adequate bearing capacity.

Safe and allowable pressures
Engineers are familiar with the technique of applying a factor of safety or load factor to an ultimate value in order to obtain a working value. The division of the *ultimate bearing capacity* by a load factor of 2·5 or 3 gives a value which is the *safe bearing capacity*. At this value of loading at the depth considered, there would be a factor of safety against a plastic shear failure.

However, foundations may also fail by suffering a settlement (and therefore a differential settlement between adjacent parts of the structure) which is more than can be permitted in the particular circumstances. It is quite likely that a factor of 3 would cover this further danger to stability if the soil is not highly compressible. On the other hand, in a highly compressible soil, the probable value of the settlement may be the controlling factor in deciding on what bearing pressure should be permitted.

When the final factor is decided upon – whether the lower figure which gives the safe bearing capacity, or a larger figure taking account of excessive settlement – the value obtained on dividing the ultimate bearing capacity by the chosen factor of safety, is called the *allowable bearing pressure*. These three terms, *ultimate bearing capacity, safe bearing capacity,* and *allowable bearing pressure* have distinct and definite meanings and must not be confused with each other or used loosely.

Ultimate bearing capacity is the pressure which causes the collapse of the foundation by plastic shear.

Safe bearing pressure refers to the pressure which can be carried without fear of plastic shear failure.

Allowable bearing pressure is the pressure which can be used without fear of deleterious settlement. Its estimation takes into account the consolidation characteristics of the soil as well as the danger of shear failure.

Effect of compressibility

A soil which is highly compressible may clearly suffer a settlement which may be unacceptable. In this instance the question of plastic shear failure does not arise, and a high load factor or factor of safety is probably required for stability of the foundation. The load factors which should be used for soils of medium strength and compressibility are not quite so easily determined. For clay, however, Skempton developed, in tabular form, a scheme relating the factor which should be applied when account is taken of the coefficient of compressibility, the shear strength, the permissible settlement, and the breadth of the footing. Fig. 10.9 is a graphical representation of his findings.

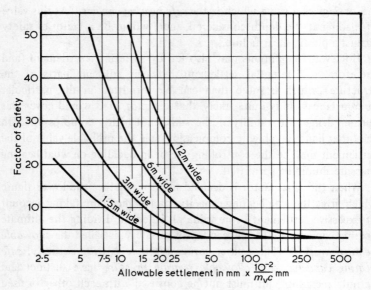

Fig. 10.9 Factors of safety for cohesive soils

The significance of the coefficient of compressibility is discussed elsewhere, but the method of using Fig. 10.9 is as follows.

(i) Determine the strength c and coefficient of compressibility m_v, or estimate their product from the state of the clay – normally or over-consolidated. Determine the reciprocal of the product.

(ii) Decide on the permissible maximum settlement in mm. This depends, to some extent, on the differential settlement likely to be encountered. Multiply (i) and (ii) together.

(iii) From the known breadth of the footing read off from Fig. 10.9 the advisable factor of safety to be used to determine the *allowable bearing pressure* from the value of the *ultimate bearing capacity*.

10.5 PILED FOUNDATIONS

When the soil on which an ordinary shallow foundation would rest is of insufficient strength to sustain the load without causing shear failure or excessive settlement, it may be necessary to use a piled foundation, the object of which is to transfer the load to stronger material at a lower level. The first problem facing the designer of piled foundations is the determination of the safe load which can be carried by individual piles or by groups of piles. The total load to be supported must be carried in such a way that the distribution of load over the area of subsoil is regulated to suit the properties of the supporting stratum. The application of the principles of soil mechanics is of great value in estimating the safe load on piles, but must be supplemented by loading tests and empirical methods based on practical experience.

Types of piled foundation
A description of the various kinds of piles used in civil engineering work is outside the scope of this book, but in studying the behaviour of a piled foundation the displacement of the soil by the pile should be considered. For this purpose piles may be classified thus:

(i) *Large displacement piles,* e.g., solid piles of timber or concrete which displace a volume of soil as they are driven and therefore compress the surrounding soil;

233

(ii) *Small-displacement piles,* e.g., rolled steel sections, open-ended tubes and screw piles, for which the displacement is relatively small;

(iii) *Non-displacement piles,* e.g., *in situ* bored piles, which are formed by boring, sinking an open-ended casing, removing the material from inside the casing and afterwards filling with concrete. Piles thus formed can be regarded as being of the nature of small piers. The surrounding soil is much less disturbed than by types (i) and (ii).

The resistance of a loaded pile is made up of two components: the toe or point resistance and the shaft resistance due to friction or cohesion on the lateral surface of the pile.

In practice piles are often described as:

(i) *end-bearing piles,* in which the major part of the load is transmitted to a hard stratum;

(ii) *frictional or floating piles,* in which the toe resistance is unimportant, the main support being due to friction and cohesion.

End-bearing piles may be further subdivided into:

(i) piles bearing on rock (Fig. 10.10a);

(ii) piles transmitting load to a relatively strong stratum of soil.

The concept of the bulb of pressure 10.14 is useful in judging the effectiveness of a pile foundation, indicating the zone which is appreciably stressed by the load and in which most of the settlement will take place.

For example, a supporting stratum of highly frictional gravel may overlie a stratum of cohesive soil at a greater depth. Whether the pressure from the piles is sufficient to cause consolidation and settlement of the underlying cohesive layer depends on several factors, the most important being the intensity of pressure applied by the foundation to the compressible layer. This pressure can be determined, and its magnitude gives some indication of the consolidation and settlement which can occur. Another factor is the pre-consolidation pressure to which the clay layer has been subjected in its natural state. This can be found from oedometer tests. If the applied pressure transmitted to the clay stratum does not exceed this pre-consolidation pressure, little settlement is likely to occur.

The pile may pass through a clay layer which is not relied upon for the sustaining of any load. The state of consolidation of such a layer should also be investigated, for should it consolidate later it would cause a downward drag on the pile and apply a load not allowed for in the foundation design.

Friction piles are used to transfer the load through a relatively weak stratum to firmer clay as in Fig. 10.10c. In normally consolidated clays of great thickness, piles are often used to take advantage of the increase in shear strength with depth. It is important that the piles should be relatively long – at least equal to the breadth of the base of the structure – in order to transfer the bulb of pressure to soil of considerably greater strength. The stability of the block of soil held between the piles, as a whole, must also be taken into consideration.

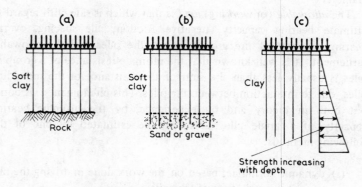

Fig. 10.10 Types of pile foundation

Friction piles in cohesionless soil. The chief function of piles in foundations of this type is to increase the shear strength of the soil by compaction. The vibration caused by driving is sufficient to compact the sand, and when it has been so compacted the piles and sand act together as one block. Vibration of loose sand by pile driving, although effective in strengthening the foundation, is liable to have the effect of causing a contraction of similar material nearby and possible settlement of adjacent structures. Bulk density measurements of the sand should be taken and shear tests made at original and compacted densities.

Bearing capacity of a pile

As for ordinary foundations, precautions have to be taken against two types of failure: shear failure of the soil and excessive settlement. As regards the first it is necessary to know the ultimate bearing resistance of the piles and to apply a suitable factor of safety. The second requirement is more difficult to assess. It calls for a knowledge of the soil profile beneath the foundation, the consolidation properties of the soil strata and the way in which these strata are stressed by the action of the piles.

The *ultimate bearing resistance* is the load at which the pile begins to penetrate into the ground without increase of load. It is sometimes specified that the ultimate load on a pile is that which causes it to settle an amount equal to or greater than 10 per cent of the pile diameter.

The *allowable* (or *working*) *load* is that which is safe with regard to ultimate bearing capacity, negative friction, pile spacing, overall bearing capacity of the ground below the piles and the allowable settlement. It is well known that the ultimate resistance of a group of piles is usually less than the sum of the resistance of the individual piles, due to interaction between the piles. This phenomenon, is borne out both in theory and by experience, by the ultimate bearing capacity of a single pile. This may be estimated by one of the following methods:

(i)   dynamic formulae, based on the work done in driving the pile against the resistance of the soil;
(ii) penetration tests;
(iii) loading tests on piles *in situ*;
(iv) static formulae, based on the principles of soil mechanics.

Dynamic formulae

The safe load on a single pile is often estimated from dynamic pile-driving formulae, which are derived by equating the energy applied in driving the pile to the work done in causing penetration. Allowances are made for the losses of energy which are known to occur but whose magnitude can only be estimated. The properties of the soil are quite inadequately represented by the measured penetration or *set* of the pile under the hammer blow. It cannot be said that these formulae are reliable or consistent in estimating the possible loading on a pile; wide variations and much uncertainty occur.

The only important difference between the many dynamic formulae is in the classification and magnitude of the allowances made for energy losses. The useful work done by the impact of the hammer blow is represented by the product of the set or penetration of the pile and the resistance or force required to cause penetration. Of this quantity only the set can be measured, and the resistance must be calculated by deducting from the applied driving energy an amount representing the various losses.

These losses are generally agreed to be caused by:

  (i) impact;
  (ii) compression of the driving cap or dolly;
  (iii) compression of the pile;
  (iv) deformation of the soil.

Some piling formulae do not take into account all these causes of energy loss. For example, the simpler formulae which neglect the weight of the pile give misleading values of the resistance when the applied driving energy is small. Other formulae do not account for the energy absorbed by the pile; thus on a heavy pile driven with a light hammer the set is quite small and the calculated value of the load-bearing resistance may be false.

A 'complete' formula is of the form:

$$\text{Applied energy} = \text{useful work} + \text{loss in impact} +$$
$$\text{loss in pile cap} + \text{loss in pile} + \text{loss in soil}$$

and this type may be used when driving in cohesionless soil.

An example of this is the Hiley formula, which considers many possible losses and gives estimated loads which agree fairly well with tested results. No formula depending on measured set can, however, be trusted to give an adequate estimation of the safe load on a pile in cohesive soil. The resistance of cohesive soils to a dynamic blow gives no indication of its resistance to static loading, and certainly no information on possible settlement.

In general, dynamic pile-driving formulae are unreliable unless their estimations can be examined in the light of past experience on similar sites or in conjunction with the results of carefully conducted loading tests. The detailed study of dynamic pile-driving formulae is outside the scope of soil mechanics.

Penetration tests

The Dutch *deep-sounding test* is designed to separate the two components of resistance by driving a cone at the end of a rod which passes through a steel tube. The *standard penetration test* is sometimes used to find the bearing capacity of cohesionless soil. Penetration tests are fully discussed in the chapter on *in situ* testing.

Loading tests

These tests form the most reliable method of ascertaining the ultimate bearing capacity of piles. They also yield valuable information about the settlement which may be expected. Often settlement rather than the ultimate bearing capacity is the deciding factor in determining the working load.

Loading tests may take either of two forms:

(i) *Maintained load method.* The load is applied by equal increments of load, the load at each stage being maintained until all observable settlement has ceased.

(ii) *Constant rate-of-penetration method.* The pile is forced into the ground at a uniform rate of settlement and the resistance continually measured. In spite of its designation this test is in principle a loading test on an actual pile, as distinct from the penetration tests mentioned above.

Details of loading test procedure are given in the chapter on *in situ* testing.

Pulling tests are sometimes carried out as a means of estimating the frictional resistance separately from the toe resistance. It is very questionable whether the resistance to pulling is the same as the static resistance to a downward load.

Settlement of individual piles

It must be borne in mind that, even under a load well below the ultimate, the settlement of a pile may be excessive for the conditions under which it is to be used. The most reliable way of checking this is by a full-scale loading test, but a useful indication of the probable settlement can be obtained by the use of the settlement influence factors derived by Poulos and Davis.

Piles with lateral loading

When vertical piles are subjected to horizontal forces, their stability against lateral movement must be assured. Resistance to lateral movement may be provided by the bearing capacity of the top strata of the ground, but if this is insufficient the resistance may be increased by connecting the pile caps by horizontal beams, or by the provision of raking piles.

In assessing the natural resistance of the soil to lateral loading it is the upper 3 or 4 m of soil which are of the greatest importance. The seasonal variation of moisture content may have an important effect. Also, the cumulative consolidation effect of lateral pressure must be considered. The magnitude of the lateral movement which takes place under transverse loading is of primary importance, as this may reduce the shaft resistance of the pile to vertical loading as well as causing eccentricity. Lateral movement may also be induced by direct loading of the soil near the piles. For example, in a piled bridge abutment there may be considerable lateral thrust on the piles due to the weight of the approach embankment.

Pile loads from static pressures

The load (W) to be carried by the pile must be augmented by the weight of the pile itself ($W + P$) to obtain the total load to be resisted. The resistance provided by the pile is in two parts: R_B, the resistance of the end bearing of the base of the pile (A_B) and R_S, the frictional resistance of the shaft against the soil. These two together must be equivalent to the load ($W + P$) multiplied by the chosen factor of safety, F. If f is the mean contact stress which can be developed along the surface of the pile acting against the soil, and A_s the surface area of the shaft, then

$$(W + P)F = R_B + R_S = qA_B + fA_S.$$

To evaluate this for a particular set of conditions, A_B and A_s are known from the dimensions of the pile chosen. The effective length of the pile (l) is that length of the penetration which is in full contact with the soil. Passage through made ground, for example, might not be considered as part of the effective length. The value of q can be obtained from Terzaghi's equation:

$$q = 1 \cdot 3 c N_c + \gamma l N_q + 0 \cdot 3 \gamma \alpha N_\gamma$$

where d is the diameter or horizontal dimension of the pile section.

The final term of this expression is small in comparison with the second because of the high value of the ratio of l to d. The third term can, therefore, be omitted. The first term does not apply to cohesionless soil but must be used for $c - \phi$ soils. The value of N_c and N_q can be obtained from Meyerhof's 'deep' coefficients (Fig. 10.4), but the values of N_q are also published by Berezantsev, Khristoforov and Golubkof in the Proceedings of the 5th International Conference on Soil Mechanics. These appear to be in close agreement with experimental findings.

The value of f (the supporting resistance along the face of the pile) is composed partly of cohesion and partly of friction. A fraction of the ultimate cohesion can be mobilized, and the frictional portion is called into play by the earth pressure acting against the pile shaft (zero at the top). Fraction is recorded as α.

K is an earth pressure coefficient less than unity, and equal to $K\gamma l$ at the bottom). The angle of friction against the pile surface is δ and the *mean* frictional resistance per unit area of the shaft is $\frac{1}{2}K\gamma l \tan \delta$. The total resistance can be calculated by multiplying by A_s which is the perimeter of the pile multiplied by the effective length. The cohesive and frictional resistance along the length of the pile shaft is, thus:

$$(\alpha c + \tfrac{1}{2}K\gamma l \tan \delta)A_s.$$

If the reasonable assumption is made that the weight of the soil displaced by the pile is equal to the weight of the pile itself, these two loads can be removed from the equation. The load then becomes W instead of $(W + P)$, and the second term for bearing pressure becomes

$$\gamma l(N_q - 1).$$

The final expression for the bearing power of the pile, if assessed from static conditions, is

$$WF = [1 \cdot 3cN_c + \gamma l(N_q - 1)]A_B + (\alpha c_u + \tfrac{1}{2}K\gamma l \tan \delta)A_S.$$

Some of the values which can reasonably be used are

$N_c = 9$: N_q, see above: $\alpha = 0 \cdot 3$ to $0 \cdot 6$ according to degree of softening. K = at least K_0, say $0 \cdot 5$: $\delta = \frac{2}{3}\phi$ to $\frac{3}{4}\phi$

but it must be remembered that this type of calculation in soil mechanics can be looked on as only a guide to the performance of the pile and cannot be a definitive and exact value.

In stiff clay ($c > 50\,\text{kN m}^{-2}$), a cavity may be formed during driving which is not so quickly closed up as with softer material, and a more conservative estimate of c should be made.

In sensitive clays the remoulding of the soil caused by pile-driving often causes a considerable decrease in the adhesion, but there is generally some recovery of strength as time passes after driving.

As the soil surrounding the pile is in a disturbed condition, the drained rather than the undrained strength would appear to be a more likely criterion for the adhesion. On this assumption the expression developed for cohesionless soil would use a value of K of about 0.7 and take δ as ϕ', the drained angle of shearing resistance.

Bored piles with under-reamed bases

In cohesive soils, holes of large diameter may be bored, cages of reinforcement fixed in place, and the whole filled with concrete. On such piles the load is carried partly by the cohesion mobilized on the surface of the shaft, and partly by the bearing pressure exerted by the base of the pile. Various patent methods of making cast-*in-situ* piles have been in use for many years. In order to study the application of soil mechanics methods to the design of such piles, the example of the under-reamed type gives the best opportunity. In this pile, the base is opened out by special tools to a conical shape, giving a wider diameter than that of the shaft. It is usually considered that the effective length of the pile should be at least four times the diameter of the expanded base, to justify the method.

Although the calculations for such a pile are not complex, it is necessary to known how much of the ultimate cohesive strength of the soil is mobilized as resistance along the surface of the shaft. The values of the load factors to be used for safety are also in question. At the Symposium on Large Bored Piles organized by the Institution of Civil Engineers in 1966, the authors Burland, Butler and Dunican reported the results of observations and test on large piles and suggested an answer to these queries. Assumptions made are:

(i) The cohesive resistance developed between the shaft of the pile and the surrounding clay should be taken as $0.3c$ ($\alpha = 0.3$).

(ii) The load factor for the load on the whole pile should not be less than 2.

(iii) The load factor for the load on the base of the pile should not be less than 3.

(iv) Effect of *overburden* in increasing bearing capacity is balanced by the *weight of the pile,* so neither appears in equations.

Using these values, the resistance of shaft and base can be developed. See Fig. 10.11 and Notation for the significance of symbols.

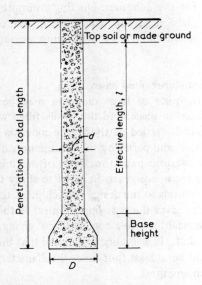

Fig. 10.11 Dimensions of bored piles with under-ream.

Resistance of the shaft developed from cohesion on its cylindrical surface ($l\pi d$):

$$R_S = l\pi d(0\cdot3c) = 0\cdot3\pi cld.$$

This value is both the ultimate and the working value. Even in the working condition, the full shaft load is developed, since it requires only a small movement for this to take place.

Resistance of the base, developed from bearing capacity (ultimate)

$$R_B = \frac{\pi D^2}{4}\,cN_c = \frac{\pi D^2}{4} \times 9c = 2\cdot25\pi cD^2.$$

Resistance of the base (working load)
= load to be supported less working load on shaft

$$= W - 0{\cdot}3\pi cld.$$

Using the Load Factors suggested, and these basic values:
Ultimate load on the whole pile = twice the load to be supported

$$2{\cdot}25\pi cD^2 + 0{\cdot}3\pi cld = 2W. \tag{1}$$

Ultimate load on the base = three times working load on the base

$$2{\cdot}25\pi cD^2 = 3(W - 0{\cdot}3\pi cld). \tag{2}$$

Simultaneous solution of equations 1 and 2 gives

$$l = \frac{W}{0{\cdot}6\pi cd}.$$

Substitution of the value of *cld* in Equation 2, gives

$$D^2 = \frac{1{\cdot}5W}{2{\cdot}25\pi c} \quad \text{or} \quad D^2 = 0{\cdot}4ld.$$

Thus, if two of the parameters are known, the other two can be calculated. Usually the diameter of the shaft is selected, and the cohesion measured. From these, a trial estimation of the effective length of the pile and the diameter of the base can be made.

It should be noted that *l* is the *effective length*. Any part of the whole length and penetration of the pile which does not provide cohesion on the surface of the shaft is not included in the effective length. Depth of made ground above the clay, and the height of the conical base are both subtracted from the total length of the pile to give the value of the effective length.

Settlement of bored piles

The settlement of a bored pile is almost entirely dependent on the settlement of the under-reamed base; the shaft contributes little after the load is applied. Settlement appears to bear much the same relationship to the diameter of the base as the working load bears to the ultimate load on the base. Equivalence is obtained in normal circumstances by the use of an empirical coefficient which, it is suggested, can be in the region of 0·02. When compared with measured settlements, this relationship gives some indication of the order of settlement, but accuracy is certainly not likely to be close. Thus:

$$\frac{\text{Settlement}}{\text{Base diameter}} = \frac{S}{D} = \frac{\text{working load on base}}{\text{ultimate load on base}} \times 0\cdot02$$

$$S = 0\cdot02\,D\,\frac{W - 0\cdot3\pi dlc}{2\cdot25\pi cD^2}$$

$$S = \frac{W - 0\cdot3\pi cld}{112\cdot5\pi cD}$$

If the settlement should approximate to zero, then

$$W = 0\cdot3\pi cld$$

or

$$l = \frac{W}{0\cdot3\pi cld}$$

Ground anchors and tension piling
Although it is a common conception that a foundation supports the weight of a structure against the action of gravity, there are many instances when the foundation must exert the reverse capacity – that of being able to hold the structure down. In single-storey flat-roofed buildings of wide extent (such as the so-called hyper-markets), the

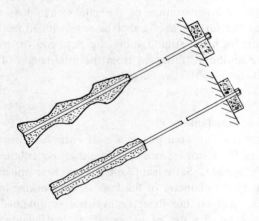

Fig. 10.12 Ground anchors

upward suction applied on the roof by negative wind pressure requires the holding down of the stanchions by heavy foundations, or by anchors. Large towers and aerial masts and their stays also require the services of anchored foundations. Tanks, built into the ground, experience an uplift under conditions of a rising water table. They must be balanced against this uplift by means of excess weight, or by some type of anchor. These *ground anchors* can also strengthen a weak soil standing to an unstable face, by holding back the surface layers against more solid materials behind. Ground anchors may be adapted to exerting a restraining pressure on the face of the soil either through relatively small anchor plates or through a revetment placed against the unstable face.

The ground anchor is formed by drilling a hole of sufficient diameter to carry the rod or strands of the tensile shaft of the anchor. This hole is cased as would be done in a borehole in unstable ground. Towards the end of the shaft the soil is reamed out by under-reaming tools as is done for a large-diameter bored pile. Under-reaming is often repeated at several points along the length of the shaft, forming *bells* whose friction against the strong undisturbed soil gives a reaction to the tension in the anchor shaft. After insertion of the rod or strands, the whole anchor is filled with grout injected under pressure. The casing is gradually withdrawn as grouting proceeds.

Post-tensioning is carried out in the usual way. Littlejohn suggests that the anchor should first be tested to 80 per cent of the ultimate tensile strength of the cable. The load should be held for five minutes before being reduced to zero. This test loading is followed by a load of the-working-load-plus-10 per cent. (The extension of the cable in these conditions should agree closely with the probable extension of the cable in free conditions). Each 24 hours the anchor load should be checked and adjusted until creep ceases. The safe working load is 5/8 of the load which produces no creep after 24 hours. Protection against corrosion must be afforded to permanent anchors, although many are intended to be temporary, and this protection can be provided in various ways with bitumen, grease and polypropylene.

The anchor may be long in relation to the diameter of the bulbs formed by under-reaming, or it may be short, the relationship of the width of the anchoring bulbs being relatively large in relation to the embedment. The danger for both of these is that they may be overstressed and fail either by the formation of a reverse borehole of the diameter of the under-reamed portion of the anchor, or by failing on inclined slip surfaces, resulting in a conical plug separating from the parent soil. Testing to failure should be carried out more often than is done at present, in order to amass information on the ultimate behaviour of ground anchors.

The linking of the performance of ground anchors with a theoretical assessment of their effects has not yet reached the textbook stage of confirmed reliability, but much discussion and experiment continues. In the meantime, on empirical evidence, ground anchors are confidently used to an increasing extent in civil engineering construction, and prove to be an acceptable means of providing an effective tensile restraint.

10.6 BEARING CAPACITY OF A PILE GROUP

It has been pointed out that for a group of piles, especially if they are friction piles, the safe load is generally less than the working load for a single pile multiplied by the number of piles. This is due to interaction between the piles and to the stability as regards bearing capacity and settlement of the block of soil in which the group of piles is driven. Another factor is uneven distribution of the total load among the piles. Even if the applied loading is uniformly distributed over the foundation slab it is unlikely that the piles will carry equal shares. The outer piles tend to be more heavily loaded than the inner ones. An exception is sometimes found in a group of piles in cohesive soil where the material is strengthened by compaction, and then the interaction of the piles may possibly increase the total resistance.

Stability of foundation block
An even more important cause of the reduction in the bearing capacity of the pile group is the settlement of the block of soil containing the piles. Referring to Fig. 10.13, it is assumed that the piles and the soil held between them act as one body in transferring the load to a deeper level and in resisting part of it by friction along the perimeter of the block of soil. The load applied to the soil at the base of the piles is equal to the load applied to the foundation less the shear resistance round the perimeter of the block.

Using the symbols given in the Notation,

$$r_p A_1 = q A_1 - LPc$$

or

$$r_p = q - \frac{LPc}{A_1}.$$

The value of r_p must be within the safe bearing value for the soil at depth z. The value of the piles lies in the reduction which they effect in the intensity of pressure on the soil and in the fact that they apply this pressure at a depth z where the bearing capacity is probably greater than at the surface.

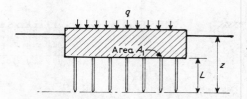

Fig. 10.13 Foundation supported by a group of piles.

To make it economically worthwhile to use a floating pile foundation in weak cohesive soil, the reduction term LPc/A_1 should be large. This term increases with increase in length of pile, cohesive resistance of the soil, and perimeter of the foundation. It also increased with decrease in area of the foundation. Foundations of this type should therefore use long piles and high values of the ratio of perimeter to area.

The value of c to be used in the above expression is a matter for conjecture. It is often taken as being equal to or a little less than the value of cohesion as obtained by the usual tests. The remoulding which occurs when piles are driven into cohesive soil often alters considerably the properties of the material. Some idea of the altered cohesive resistance can be obtained by measuring the force required to withdraw the pile. The time effect is important: the resistance of the pile immediately after driving may be much less than it becomes some time after driving has ceased.

Settlement criterion

The action of a group of piles and the significance of the ratio *length of pile/breadth of supported foundation* will be appreciated by referring back to Fig. 6.9, which shows that the pressure bulb for a wide foundation is larger than for a narrow footing and may reach strata well below the level at which the smaller pressure bulb ceases to have effect.

Under a friction pile foundation a similar bulb of pressure is developed, and the width of this bulb may be considered to represent the effective area of soil supporting the load. Under a single pile this area is a much larger multiple of the area of the foundation than is the bulb under a group of piles (Fig. 10.14).

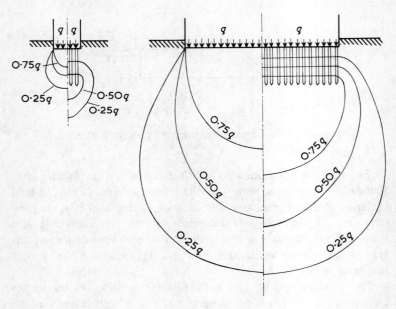

Fig. 10.14 Pressure bulbs under piled foundations

The safe load for a given settlement as estimated for a single pile thus bears little relation to the load/settlement relation for a group. Further, along the length of a single friction pile the load may be well distributed, and the highest stress developed in the soil may be less than that caused under an ordinary footing.

The highest stress is also applied at a lower level where the soil is likely to have a higher resistance. Under a wider foundation, where the ratio of length of pile to foundation width is smaller, the highest stress developed under the group of piles does not differ much from that produced under a normal footing. Fig. 10.14 shows that when the piles are long in relation to the width of the foundation the bulb is carried down to a lower level where the soil is more capable of carrying the stresses applied. When the piles are short in relation to the width of the foundation, the presence of the piles have little effect in altering the soil stress at any selected point below the foundation.

Safe bearing capacity of a pile group

The safe bearing capacity of a group of *n* piles is sometimes expressed as *βnR,* where *β* is a factor which depends on the soil properties, the number of piles, and the ratio of spacing to pile diameter. The factor *β* is sometimes called the efficiency of the group. Several empirical formulae have been devised to estimate the efficiency of pile groups, but there are too many variables for these to be really reliable.

Experiments carried out at the Building Research Station on square groups of model piles in cohesive soil show that for a given length and number of piles there is, for each group, a certain spacing at which the type of failure changes. For spacings closer than this, block failure occurs; for wider spacings, failure occurs in individual piles. These experiments are reported in a paper by Whitaker.

Piling is not always the solution for poor foundation material. Highly compressible and weak soil can be made to support an external load by the expedient of first relieving it of the pressure applied by the strata above it.

To effect *total compensation,* a rigid box foundation is constructed, extending to such a depth that the material excavated is equal in weight to the total weight of the building. If the estimate of the weight of the building has been accurate, little or no relative movement between the building and the surrounding soil should occur. In practice, total compensation is almost impossible to achieve, and partial compensation has to be accepted. A proportion of the weight of the building is cancelled by excavation and some settlement allowed. This settlement is caused by the excess weight defined by the difference between the weight of the building and the weight of the excavated soil. Even in such instances settlement can be stopped by the use of such devices as *control piles* devised by Gonzalez Flores and used in buildings in Mexico City. Piles are passed through the base of the buoyant foundation and the excess weight carried by high-tensile steel wires between a yoke on the control pile and the structure of the buoyant foundation. The wires are in the numbers required to carry the excess weight at the limit of elasticity. They are shared between control piles distributed over the base of the foundation. Such costly remedial measures are not likely to be required unless the soil is highly compressible, as it is in Mexico City.

Buoyancy foundations are costly since reinforced concrete cellular construction is necessary to keep down the dead weight and to spread the load evenly. Excavation also is costly and difficult and causes disturbance of the soil. Trouble may be experienced from swelling of the soil on removal of the overburden pressure by excavation. A variation of the compensation method is to construct a basement in the centre of an otherwise uniformly-loaded foundation. This has the effect of compensating for the dish-shaped settlement normally caused by uniform loading on a compressible soil. A more uniform settlement over the whole area of the building is produced.

APPLICATION OF THE WORK OF CHAPTER 10

Application 10A
If a pile is expected to carry 550 kN and the diameter selected is 0·5 m, the expressions developed above can be used to obtain the values of the length and of the diameter of the base. The cohesive strength of the clay is found to be 86 kN m^{-2} .

Effective length of the pile in this clay should be at least

$$l = \frac{W}{0 \cdot 6 \pi c d} = \frac{550}{0 \cdot 6 \times \pi \times 86 \times 0 \cdot 5} = 6 \cdot 8 \text{ m.}$$

Diameter of base, on this assumption:

$$D^2 - 0 \cdot 4 \, ld = 0 \cdot 4 \times 6 \cdot 8 \times 0 \cdot 5 = 1 \cdot 36; \quad D = 1 \cdot 2 \text{ m.}$$

Thus the design would be

$$l = 6 \cdot 8 \text{ m}; \quad d = 0 \cdot 5 \text{ m}; \quad D = 1 \cdot 2 \text{ m.}$$

These figures should be accepted as a guide. They are not likely to be more exact than within 10 to 15 per cent of the 'correct' figure.

In this problem, the conditions of settlement are:

$$S = \frac{W - 0 \cdot 3 \pi c l d}{112 \cdot 5 \pi c D} = \frac{550 - 276}{36479}$$

$$= 0 \cdot 0075 \text{ m} = 7 \cdot 5 \text{ mm.}$$

If the settlement should approximate to zero:

$$l = \frac{550}{0 \cdot 3\pi \times 86 \times 0 \cdot 5} = 13 \cdot 6 \text{ m}$$

to obtain negligible settlement.

If the effective length is kept at 6·8 m, the indications are that the settlement would be in the region of 8 mm.

Application 10B

An extract from the result of a site investigation is given below. It is proposed to found a circular shaft which will impose a load of 6700 kN on its foundation. The shaft is 2·7 m diameter. For a shaft depth of 7 m, to what size should the foundation be constructed? G.L. is 273 m above ordnance datum.

Borehole T/732

Level A.O.D. (m)	Type of sample	Density of sample (kg m^{-3})	Cohesion c (kN m^{-2})	Angle of shearing resistance
268·5	Sandy clay	2205	33	3°
265·5	Boulder clay	2183	165	3°
262·5	Boulder clay	2235	138	7°
255·5	Boulder clay	2300	165	4°

The value of the coefficient of compressibility (m_v) is $0 \cdot 012$ m^2 MN^{-1} at a pressure of 80 kN m^{-2} and $0 \cdot 0088$ m^2 MN^{-1} at a pressure of 1500 kN m^{-2}.

The settlement of the shaft should be restricted to 10 mm.

(i) There is a weak clay at 4·5 m below the surface, so this is likely to be a 'deep' foundation. Use Fig. 10.4 in obtaining *bearing capacity factors* which will be used in the equation

$$q = cN_c + \gamma z N_q.$$

The third term of Terzaghi's equation does not appear this time, as the values of ϕ are too small to bring in N_γ.

(ii) *Ultimate bearing capacity.* The calculations should be carried

Table 10.1 Calculation of bearing capacity (Application 10B)

(1) Level A.O.D. (m)	(2) Depth, z (m)	(3) Bulk density γ (kg m⁻³) (kN m⁻³)	(4) γz (kN m⁻²)	(5) φ	(6) Nq	(7) Nc	(8) c (kN m⁻²)	(9) cNc (kN m⁻²)	(10) 1·3 × cNc (kN m⁻²)	(11) γzNq (kN m⁻²)	(12) (10)+(11) Ultimate bearing capacity (kN m⁻²)	(13) Allowable bearing pressure Divide by E = 3	(14) Add γz
268·5	4·5	2205 21·6	97	3°	1·6	12	33	395	514	155	670	220	320
265·5	7·5	2183 21·4	161	3°	1·6	12	165	1980	2574	258	2830	940	1100
262·5	10·5	2235 21·9	223	7°	2·7	16	138	2208	2870	602	3470	1160	1380
255·5	17·5	2300 22·6	400	4°	1·9	13	165	2145	2789	760	3550	1180	1580

Ground level = 273 A.O.D.

out for the four samples tested, and this is best done in tabular form. In a full investigation, many more than four figures will have to be obtained. Table 10.1 shows the method:

Columns 1, 2, 3, 4, 5 and 8 give the data from the site investigation. Columns 6 and 7 are evaluated from Fig. 10.4 using values of ϕ. Column 10 transforms the bearing capacity for a deep strip to that for a square footing, or a circular footing.

(iii) From the Table 10.1, the ultimate bearing capacity varies from 670 to 3550 kN m^{-2}. The final two columns, 13 and 14, cannot be completed at this stage.

(iv) Now, enter Fig. 10.9 with an abscissa equal to

$$\text{mm of allowable settlement} \times \frac{10^{-2}}{m_v c}.$$

The values of these abscissae for the two values of m_v given are: (assuming mean values of c to be 150 kN m^{-2})

$$10 \times \frac{10^{-2}}{150 m_v} = \frac{1}{1500 m_v} \text{ when } m_v \text{ is in m}^2\text{kN}^{-1}$$

$$= \frac{10^3}{1500 \times 0 \cdot 012} \text{ and } \frac{10^3}{1500 \times 0 \cdot 009} \text{ which are 56 and 74 mm.}$$

for bases up to 3 m wide. Both of these lie in the zone of a factor of safety of 3. If they had been less in value, the factor of safety would have been greater.

(v) Use factor of safety, 3, to develop column 13, then add the weight of soil displaced by the construction of the shaft to give the gross value of the permitted loading. It is best at this stage to draw a graph of the allowable bearing pressure (column 14) against depth below the surface. Read off the allowable bearing pressure for a depth of 7 m, the proposed penetration of the shaft. It is in the region of 1000 kN m^{-2}.

(vi) The area required of the foundation to the shaft is thus 6·7 m^2 which gives a square of 2·6 m side or a circle of *just under 3 m diameter.*

The required dimension is, therefore 3 m diameter or more.

The current use of electronic desk or pocket calculators may give the impression that 'accuracy' is being achieved, for the figures obtained are correct to many significant figures. It must be remembered that this correctness is based only on the data fed into the problem. In soil mechanics problems the input data are not, by any means, 'accurate' since they deal with variable site conditions not amenable to exact quantification. The figures obtained are not much more than a guide which allows the judgement of the engineer to be exercised with a degree of confidence slightly greater than if he had to rely on past experience only. Dependence on electronic calculators alone is to be avoided.

CHAPTER 11

Improvement of soil properties

In the solution of many engineering problems it is necessary to improve the properties of soil, whether as foundation material or as a material of construction in embankments, dams, and other artificial works. The requirements in these conditions are that the soil should be capable of sustaining the applied loads without serious deformation, and that it should retain its strength and stability indefinitely.

A common example of the necessity for such improvement arises in the foundation of roads and airfield runways.

A road or runway usually consists of the *pavement,* whose function is to distribute the effects of the wheel loads over a sufficient area of the *subgrade,* or natural foundation material underlying the pavement.

The pavement comprises one or more layers of artificially compounded material covered by a durable and waterproof running surface. The layers below the running surface are known as the base and sub-base.

The fundamental techniques by which the properties of natural materials can be improved are *drainage, grading, compaction* and *stabilization.*

Since the percolation of water always has a weakening effect on natural soils, a well-graded material which can be compacted to a high density, and therefore offer maximum resistance to percolation, is the most suitable for construction.

255

11.1 DRAINAGE

The importance of the moisture content of any soil used for engineering purposes has been emphasized in several of the preceding chapters. In general, the more desirable properties of a soil are mobilized and improved by a reduction in the moisture content. Conversely, an increase in moisture content is generally accompanied by deterioration of strength and bearing capacity, especially in cohesive soil.

More important, however, than its absolute value is the possible variation in moisture content. Such variation may be seasonal or caused occasionally by abnormal conditions, but whatever its cause and whatever its magnitude it is an unwelcome occurrence. Soil structures such as embankments, cuttings, and foundations are designed for soil in a specified condition, and any change from that condition must be prevented or minimized.

The maintenance of the soil in a strong and stable condition thus depends on removing excess water from the site and preventing the access of water. The drainage necessary to achieve this purpose should maintain the subsoil itself at as uniform a moisture content as possible.

The effects of varying underground strata

Although water can percolate to some extent through any soils except dense and homogeneous clays, its rate of flow in porous noncohesive soils is markedly different from that in cohesive materials. Strata can without much error be described as *pervious* and *impervious,* although 'more pervious' and 'less pervious' would be more exact definitions. Using the vivid, if inexact, terms pervious and impervious, it is possible to elucidate the behaviour of ground water for simple regimes of flow.

In Fig. 11.1a it is shown how a depth of saturated soil is built up in pervious material overlying an impervious stratum. The natural flow of the water under the hydraulic gradient, indicated by the curve of the water table, results in the appearance of springs at the junction of the two strata. Fig. 11.1b shows how the flow through a pervious stratum may be opposed by an impervious stratum moved out of position by a geological fault. The head of water developed in the pervious material may be sufficient to cause springs to appear on the line of the fault.

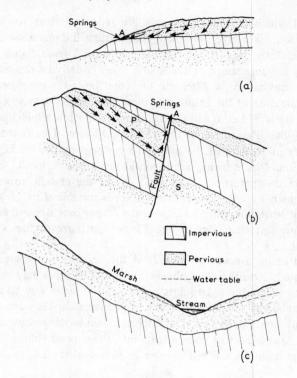

Fig. 11.1 Flow of ground water

Below stratum *P* lies an impervious stratum, and below that again there appears another waterbearing stratum or *aquifer*. The water in this may well be under pressure, and borings sunk through to this aquifer might release artesian pressure causing a mixture of sand and water to rise in the boreholes. In Fig. 11.1c the presence of marshy places and streams is shown to be evidence of the water table coming to the surface.

French drains
This is a variation of the method of draining by agricultural tile drains. A larger surface area is provided than that presented to the ground water by the agricultural pipe. The pipe is laid, with the usual open joints, in the bottom of a trench at least 300 mm wider than the diameter of the pipe. The trench is then filled with layers of graded

257

material which is coarser towards the pipe and finer towards the natural soil. Each layer, although coarser than the one above, should be sufficiently fine to prevent finer material from being washed through the interstices. If the coarse layer underlying the one to be supported is called the *filter* and the other layer the *protected layer*, an examination of the gradings of these two layers shows whether the filter is suitable. Let us call the size of particle below which lies 15 per cent of the filter material the 15 *per cent size*. This is the largest diameter of particle of the lowest 15 per cent on the grading curve, and is found by drawing a horizontal line through the 15 per cent mark on the 'percentage finer' axis to cut the grading curve. It has been shown by experiment that a filter is suitable if its 15 per cent size is at least four times as large as the 15 per cent size and not more than four times as large as the 85 per cent size of the protected material.

Such drains are successful only if the backfill is well graded and compacted, conditions not always fulfilled. The pipes may be subject to silting unless the open joints are covered in some way to allow of the granular filter forming in time a protective arch. They are useful in positions where open drains or ditches would be dangerous to traffic. French drains built as counterforts into the slipped volume of a soil slope can drain the soil and improve its shear resistance.

Counterfort drains

A variation of the French drain is the Counterfort drain. This can be used to stabilize a slip in a bank. Trenches are dug into the slope until the slip surface is encountered and taken beyond this point sufficiently to be well bedded into the stable material beyond. These trenches are then filled with graded drainage material and have the double function of removing water and forming a physical support to the bank.

Interceptor drains

If surface and subsoil water is prevented from reaching the area over which the water table must be kept at a low level, the drainage system on the site can be less elaborate. Interceptor drains at such points as *A* in Figs. 11.1a and b would trap the flow of spring water and ensure that drainage systems on the lower slopes would deal with surface and subsoil water from local areas only.

Interceptor drains round a site often run along the contours or along the summits of embankments, and may be of considerable length. Their function of preventing water from reaching the site, and so reducing the cost of the site drainage, is, however, of importance. They also fulfil the function of lowering the water table on a waterlogged site, thus allowing the construction of foundations to proceed in drier conditions.

Well drains

A thorough examination of the geological conditions may show the possibility of removing water by draining it downwards through wells or boreholes into previous strata lying at a lower level. In Fig. 11.1b, for example, a system of vertical boreholes driven through the pervious stratum P to the pervious stratum S would remove by drainage into S much of the water causing springs along the fault line. If wells are to be used in this way, the engineer must have a thorough and exact knowledge of the geology of the area. Another use of wells and boreholes is in the rapid but temporary drawdown of the water table for the purpose of building a foundation or other structure below the normal water table. Wells of large diameter or 'well-points' of small diameter are put down all round the site and pumping is started, either with separate pumps for each well or with linked pumping units for the whole system. By this means it is possible to lower the water table by as much as 5 m so long as the pumps are running. Even greater depths can be reached by multi-stage installations.

Sand drains

In a previous chapter, sand drains are discussed as a means of accelerating settlement of a clay layer before the final structure is built. They are thus a means of improving the properties of soil, by reducing the amount of settlement which would take place under working loads. This rather specialized drainage method is mentioned here for the sake of completeness, but the full discussion is in a previous chapter.

Mole drains

Clay is not amenable to drainage, for most of its water is held by capillary forces and will not move under gravity. Further, the voids in clay are extremely small and there is a high resistance to flow.

The mole drain method of lowering the water table in clay,

although previously used, developed rapidly during the Second World War with the abandonment of steam traction and the substitution of track-laying tractors. The mole plough is bullet-shaped and is mounted on the end of a flat bar which holds it at some distance below ground level. As a tractor draws the plough along, a nearly cylindrical channel is formed underground. This channel retains its shape for many years if it is not subjected to heavy loads, and will remove excess water in the clay. A 45 to 50 h.p. track-laying tractor will produce a 75 mm channel at a depth of 600 mm, and a 25– to 30–h.p. track-laying tractor will give at 70 mm mole at 450 mm depth. These mole laterals should be drawn (in the uphill direction) at about 3 to 5 yard intervals and should connect with a main drain or ditch.

Temporary stabilization as a substitute for drainage

Excavation in soft and unstable material is sometimes rendered very difficult by the ineffectiveness of normal drainage methods when applied to fine-grained soils. Such soils do not give up moisture under ordinary gravitational forces and remain fluid and unmanageable. Specialized expedients must be adopted in order to strengthen such soil temporarily in order to allow of the completion of the excavation.

These expedients may be divided into two groups:

(i) those methods which do not attempt to remove water but which strengthen the soil in order to counteract the deleterious effect of excessive moisture;
(ii) means of forcing the moisture out of a fine-grained soil.

Geotechnical processes. The stabilization of the volume of soil under a foundation area for a length of time sufficient to allow of the excavation being made is carried out by what are known as *geotechnical processes*.

The methods adopted are:

(i) freezing of the soil;
(ii) injection of cement grout;
(iii) production of a silica gel within the voids of the soil.

The obvious effect of the circulation of low-temperature brine through pipes in the soil is to *freeze* the water in the voids and

260

produce a temporarily strong material through which excavation can take place. Like other methods described in this section, freezing requires considerable plant and is an expensive expedient. Freezing can be applied to all types of soil, but is particularly effective in silts.

Injection of cement grout is successful when the voids in the material to be stabilized are sufficiently large to allow penetration. Fine materials tend to act as a filter, and even a very liquid grout cannot be pumped far into fine material. At the International Conference on Soil Mechanics of 1936, Terzaghi showed that if the effective size of a compact sand is less than 1·4 mm, or that of a loose sand less than 0·5 mm, cement grout will be filtered out.

For fine material (less than about 0·5 mm) the formation of a *silica gel* within the voids of the material produces a stabilized soil of adequate strength for the excavation of foundations. The precipitation of this gel is carried out by specialist methods and involves the pumping into the soil of sodium silicate, followed by a solution of a salt such as calcium chloride. In the fine material for which this is adapted each perforated pipe, through which the solutions are pumped, can serve a cylindrical volume of only about 600 mm diameter. The method is expensive, requires many boring pipes set close together, and its use is confined to small excavations in coarse and medium sands.

Larger volumes of porous strata may be sealed by the use of silica gel, the silicate and chloride solutions being mixed before pumping, but prevented from reacting immediately by the addition of a restrainer. By the time the reaction commences, the liquid has had time to penetrate under pressure to some distance from the injection pipe.

Electro-osmosis. It is well known that drainage of silts and clays cannot be completely accomplished by providing easy gravitational paths for flow of water. Water is held back against gravity by capillary and frictional forces in the minute voids of fine-grained material: under such conditions normal methods of drainage produce but little improvement of the soil properties. The drainage of soft silts and clays is, however, often a requisite preliminary step in a constructional scheme, and an electrical means of forcing the water out of the soil has been tried with some success both experimentally and on a large scale.

If an electric current is passed between two electrodes buried in a saturated fine-grained soil, the free water flows towards the cathode.

261

If this is made in the form of a porous and hollow metal cylinder, the extracted water can be pumped out.

Soils which exist in the natural state at a moisture content close to or over the liquid limit can be forcibly dried by this means to a state indicated by the lower ranges of the plasticity index. *L*. Casagrande showed by experiment that a fat clay at a liquid limit of 67 could be reduced by electro-osmosis to a moisture content of 25 to 40 per cent in the short period of 100 hours. The dried results were obtained by a close spacing of the electrodes.

Electro-osmosis is particularly effective in preventing rises in the floors of excavations within sheet-pile cofferdams. Very little excess hydrostatic pressure is sufficient to cause the 'boiling' of soft, nearly liquid material unless electrical drying can be adopted.

The use of hydrated and quick lime. By chemical action water is absorbed or thrown off as steam by the use of lime, and the soil is apparently 'drained' and hardened to take construction traffic.

11.2 COMPACTION

It is important that the differences between *compaction* and *consolidation* should be clearly understood.

Consolidation refers to the gradual expulsion, by continued pressure, of water from the pores of saturated cohesive material, with consequent reduction in volume.

Compaction is the packing together of soil particles with the expulsion of air only. It is accomplished by rolling, ramming, or vibration, and results in a decrease in the volume of air voids and an increase in the density of the soil.

The object of compaction is to improve the desirable qualities of the soil. Three of the properties which are of importance in the construction of roads, runways, and embankments, are:

  (i) high shear strength;
 (ii) low permeability and water absorption;
(iii) little tendency to settle under repeated loading.

In considering how compaction can be profitably applied to soil it is necessary to know:

   (i) the state of compaction of the material as it occurs naturally;

(ii) the maximum compaction which it is possible for the material to attain;

(iii) what proportion of the maximum compaction can be achieved by field equipment on the site.

Moisture-content/dry density relation

The state of compaction of any material is measured by the dry density of the material. Two figures are required for the determination of the state of compaction: the bulk density γ of the material and its natural moisture content m.

The dry density γ_d is given by

$$\gamma_d = \frac{\gamma}{1 + w}.$$

If a soil is compacted to the state when all air voids have been removed but no mosture expelled, it is said to have reached *saturation*. For any given moisture content this *saturation dry density* can be calculated if the specific gravity of the particles is known. For construction in the field this is an impossibly high standard of compaction, and some lower arbitrary density must be chosen as the *maximum state of compaction* attainable.

The relationship between dry density and moisture content for a particular compactive effort is best studied by reference to the British Standard compaction test, often known as the Proctor Test. In this test a sample of soil is compacted in a specified manner, and its density and moisture content are measured. The process is repeated for a series of different moisture contents. The results of this test are plotted, dry density against moisture content. A typical example is the lower curve in Fig. 11.2.

The state of compaction reached in the B.S. Compaction Test varies with the moisture content of the material. When the soil is wet it is possible for the standard number of blows to reduce considerably the few voids not occupied by water. The dry density reached may thus approach the saturation dry density, which, because of the presence of a large amount of moisture, is quite low (B in Fig. 11.2). On the other hand, when the moisture content is low the theoretical saturation dry density is high. In this condition no amount of compaction is sufficient to cause much reduction in the air voids, as there is insufficient moisture to provide the lubrication necessary to enable the particles to pack closer together under the blows. The final

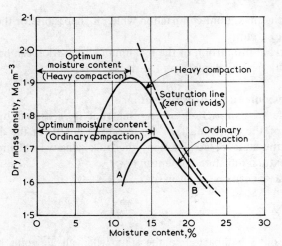

Fig. 11.2 Moisture/density relationships

dry density reached may thus be quite as low (*A* on curve in Fig. 11.2) as when the soil is wet. At some intermediate moisture content there is an optimum point at which the standard compaction results in a maximum dry density. This is known as the *optimum moisture content*. The saturation curve (dotted line in Fig. 11.2) is plotted from the equation

$$\gamma_\mathrm{d} = \frac{G_\mathrm{s}\gamma_\mathrm{w}}{1 + wG_\mathrm{s}}.$$

To simulate conditions obtainable in the field from modern heavy compacting equipment an alternative standard, known as *heavy compaction*, has been adopted. This standard is based on a method devised by the American Association of State Highway Officials. As will be seen from Fig. 11.2, the higher maximum dry density obtained occurs at a lower optimum moisture content. Thus the maximum dry density and the optimum moisture content are not fundamental properties of the soil, but depend upon the energy of the compactive effort applied. A further method of compaction is also possible under BS 1377 when a vibrating hammer is used to apply the compactive energy.

When either of these standards is adopted for a particular job the actual dry density to be obtained in the field, described as the *relative*

264

compaction, is expressed as a percentage of that obtained from the test. It is desirable that the air content should not exceed 10 per cent of the total volume, but a lower maximum value is often specified.

Effect of compaction on soil properties

As is to be expected, the shear strength of soil increased with the amount of compaction applied. The more the soil is rammed or rolled, the greater is the value of the cohesion and of the angle of shearing resistance. Comparing the shear strength with the moisture content for a given degree of compaction, it is found that the greatest shear strength is attained at a moisture content lower than the optimum for maximum density. It might be inferred from this that it would be an advantage to carry out compaction at the lower value of the moisture content, but this is not the case. A soil compacted at a condition such as that indicated by the point *A* in Fig. 11.2 will tend to take up moisture and become more nearly saturated as shown at *B*, with consequent loss of strength. A high degree of compaction, with a minimum of air voids, also reduces the tendency for settlement under steady and repeated loading.

Compaction in the field

The standard compaction tests are very useful as a guide to the moisture content at which compaction should be carried out. In translating the results of laboratory tests into practice the necessary factors are:

(i) the adjustment of the natural moisture content in the soil to the value at which field compaction is most effective;
(ii) the provision of compacting equipment suitable for work with the type of soil encountered.

In the field it is often impossible to work at the ideal moisture content. Working conditions generally have to be modified to suit the value of the natural moisture content of the soil, the conditions at the site and the prevailing weather.

If the soil is drier than that having the optimum moisture content as shown by the standard test, it may be wetted in order to bring its moisture content up to that value before compaction, or it may be compacted in the drier state by giving it a greater amount of compaction – equivalent to the use of heavy plant which often requires a lower moisture content than the standard optimum.

If the material is wetter than the optimum no amount of compaction will increase the dry density beyond that shown by the normal moisture–density curve. In dry weather, however, the natural moisture content can, if necessary, be reduced by pulverizing the soil. Whatever the amount or method of compaction used, 5 to 10 per cent of air voids is left in the soil, even where the maximum dry density is obtained.

It is common practice to specify the *relative compaction* required, referred either to the *B.S. Compaction Test* or to the *Heavy Compaction Test,* usually to a figure of 90 to 95 per cent. A range of working values of moisture content must also be specified, based on the *optimum moisture content.*

The method of specifying using the relative compaction is open to the objection that the compactive effort of field equipment (depending on the weights applied and the number of applications) does not generally conform exactly to either of the standard tests. An alternative is to specify the minimum air content, the value of which is chosen from laboratory tests. If possible, full scale tests should be carried out to ascertain the most satisfactory way of achieving the desired result.

For the air-void method it has been suggested that the selection of moisture content should be based on the equilibrium moisture content of the soil rather than on the results of the standard compaction tests. The equilibrium moisture content can usually be estimated from the natural moisture content at a depth of 1 or 1·5 m below the surface, unless the water table is high.

Compacting equipment
Methods of compaction may be divided into three categories: rollers, rammers, and vibrators. Table 11.1 shows a list of compacting equipment in general use.

For efficient compaction use should be made of high intensity large-area pressures applied to relatively thin layers of material. The type of roller equipment used and the intensity of pressure applied must, of course, be adapted to the material under compaction. The soil, for example, must possess sufficient resistance to plastic deformation to allow a compacting stress to be developed. If such resistance to deformation is not present, as, for example, in a wet clay, a high intensity of rolling stress causes large deformations, and no compaction results.

When we consider a layer of loose soil or stabilized material having perhaps a relative compaction of about 70 per cent of that obtained by the British Standard test and the object is to compact this to, say, 95 per cent relative compaction, it is natural to think of rolling so that the bottom layers are compacted first. A high intensity load is required which will penetrate through the loose material and gradually 'ride out' to the surface. This conception produced the sheepsfoot roller with tapered feet, which gives a high intensity of loading, or with club feet, which gives a greater coverage for each pass of the roller.

Table 11.1 *Compacting equipment*

	Type	*Suitable for*
Rollers	Smooth-wheel	All soil types except wet clay and uniformly graded sand
	Pneumatic-tyred	Most soil types, particularly wet cohesive soil
	Sheepsfoot	Cohesive soil, not wet
	Scrapers and heavy excavation machinery. Track-laying plant	Most soil types when no specialized compaction plant is available
Rammers	Dropping weight	Small jobs, e.g. trenches
Vibrators	Vibrating rollers	Granular soils
	Vibrating plate	Most soil types

Results of tests at the Road Research Laboratory and experience in practice have shown that, in general, the sheepsfoot roller gives comparatively poor performance, and there is no evidence that it compacts from the bottom up. Sheepsfoot rollers are, however, effective in compacting dry clays, and though not very suitable for British conditions, are quite useful in dry areas.

Further full-scale tests at the Road Research Laboratory have shown that the smooth-wheel 8 ton roller is the most generally useful compaction plant, particularly for sand and gravel-sand-clay mixtures.

Heavy pneumatic-tyred rollers appear to be more satisfactory on wet cohesive soil, and their output in quantity is relatively high. Rolling equipment may be expected to compact layers not greater than 150 mm thick when compacted (about 200 mm loose). In using roller equipment it is necessary to ascertain by test the number of passes to effect the desired degree of compaction. Generally soils with higher moisture content are compacted with fewer passes than drier soils.

An example of equipment operating on the ramming principle is the frog-rammer, which is often used for compacting fill in trenches. It can compact thicker layers of soil than other types of compacting equipment, but the moisture contents at which it operates best are somewhat low for the climatic conditions in Britain. Its output is small compared with that of rollers.

In Britain, work has been done with a vibrating plate and vibrating rollers. The plates can compact material in places where no roller could operate. These pieces of compacting plant run at 1000/1800 cycles per minute, and are available between 800 and 2000 kg. The lighter German plate compactors of 50 and 150 kg are effective in confined places.

11.3 STABILIZATION

In its widest sense the term stabilization comprises any process which increases or maintains the natural strength of the soil. In this sense it includes compaction, drainage and the stabilization of banks by sowing of grass and planting of trees. More usually, however, the term is restricted to the improvement of soil properties by treatment with additives.

A stable soil may be defined as one which:

(i) withstands, without disintegration, the small deformations produced by loading;
(ii) retains these qualities under varying weather conditions.

From the point of view of bearing capacity the best materials are those which derive their shear strength partly from friction and partly from cohesion. Thus a material which is deficient in either of these qualities can be improved by the addition of some other material which will supply the deficiency. For example, the bearing capacity of clay can be increased by the admixture of sand and gravel, while coarse-grained material has its cohesive properties improved by the addition of a binder such as cement, tar or bitumen. In fact this principle of mixing a well-graded aggregate with a binder is the foundation of all methods of road construction from the original water-bound macadam road to the most modern surfaces and foundations.

Again, assuming that the desired strength has thus been attained, this strength must be permanently maintained. The chief agencies tending to change the state of the material and reduce its strength are rain and frost. There are several methods by which some assurance can be reached that the compacted material will substantially retain its original moisture content and strength:

(i) prevention of surface water from entering the soil by an impervious upper layer (e.g. concrete or asphalt);

(ii) removal of surface water and some, at least, of the subsoil water by an adequate system of drainage;

(iii) use of material whose strength is dependent mainly on frictional resistance and is unaffected by alteration in moisture content.

The processes of stabilization may therefore be grouped under two headings:

(i) the raising of the bearing capacity of the material by the addition of aggregate or binder;

(ii) the preserving of the stability of the material by the use of water-proofing techniques.

Some methods of stabilization, e.g. the use of bitumen, are really a combination of these two.

Mechanical stabilization

Since the earliest days of road-making, a traditional technique has been to use a well-graded granular material containing sufficient fines to retain moisture and so provide the necessary cohesion. When compacted, this material is very stable. In some parts of the world including several areas of Britain, a natural soil is found of the right mixture and complying with the requirements for stability. When such material is used for road construction only compaction is necessary to

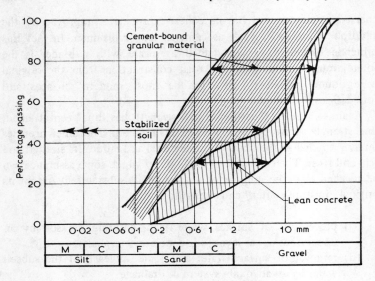

Fig. 11.3. Gradings for soil–cement construction

produce a reliable base. The limits within which soils are mechanically stable is very wide and almost any naturally occurring soil can be used if mixed so that the proportions of granular material and binder produce a high dry density under the British Standard Compaction Test.

Cement stabilization

This term is applied to various materials in which natural soil, washed or processed granular materials, or waste products such as slag or fuel ash, are improved by the admixture of a small proportion of Portland cement, generally not more than 10 or 15 per cent. The addition of cement develops a higher shearing strength and a greater resistance to the ingress of water.

The chief use to which cement-stabilized material has been put in many countries is in the construction of sub-bases and bases for roads, but it has also been used for temporary or lightly trafficked surface layers.

The term *soil-cement* is generally applied to natural soil into which the cement is mixed *in situ*. It was first widely employed in the Second World War for the inexpensive and rapid construction of roads and airfields. After the War, many lightly trafficked roads in housing estates were constructed by the mix-in-place method. One of the difficulties encountered has been that machinery used for mixing the cement into the soil in place could not effectively penetrate more than 150 to 180 mm. Thus for a deeper base or sub-base, two-course operation had to be employed, or other and more costly means of mixing used. More powerful machinery is now available capable of penetratin, by the mix-in-place method to a depth of up to 300 mm.

With the increase of traffic, however, the application of cement has been extended to other materials giving greater strength.

Soil-cement thus forms the first link in a chain leading from soil to concrete. The next stage beyond the *in situ* stabilization of natural soil is *cement-bound granular material* (CBGM) which uses natural granular material with a limited percentage of fines. The material is mixed with cement mechanically, and developes a somewhat higher strength than the simple stabilized soil.

The next stage towards ordinary concrete is known as *lean concrete* which is widely used for bases for roads, and, for a given strength, requires less cement than normal soil-cement. The gradings required for the three classes of cement stabilization are shown in Fig. 11.3. Of the three classes this volume is concerned only with soil-cement.

Soil suitable for this form of stabilization should be well-graded, with a coefficient of uniformity not less than 5 per cent. The liquid limit should not exceed 45 and the plastic limit should be less than 20. An increase in clay content not only increases the difficulty of thorough mixing with the cement, but also reduced the maximum dry density obtainable and therefore increases the proportion of cement required to attain a specified strength. The soil should be free from constituents such as organic matter or sulphates, which would have a deleterious effect on the setting of the cement.

A minimum strength of $2.8 \, \text{MN m}^{-2}$ at seven days is generally specified, as determined by the unconfined compression test. Alternatively, a strength of $3.5 \, \text{MN m}^{-2}$ for cubes may be used. For soil cement there is little advantage in having the strength much above these figures. With strength in excess of $5 \, \text{MN m}^{-2}$ there is a likelihood of cracks occurring due to shrinkage; also, the quantity of cement required to develop such strengths would be uneconomic.

Assuming that the soil is suitable for cement stabilization, the first step is to decide on the necessary moisture content and dry density. A small increase in density makes a relatively large increase in strength for the same cement content. The dry density must therefore be as high as practicable. The quantity of water to be added thus depends on the original moisture content of the soil to be stabilized, the total quantity being sufficient to achieve the desired degree of compaction and to hydrate the cement. The B.S. heavy compaction is a good indication of the necessary condition, and the air voids should be limited to 5 per cent. In the compaction test a trial mixture containing, say, 10 per cent of cement (the percentage is reckoned on the weight of dry soil). Then unconfined compression tests are carried out at this density and moisture content, using say, 5, 10, and 15 per cent cement, which enable the most suitable proportion of cement to be determined. The procedure in the field consists of pulverization of the soil, mixing with cement, watering to the desired water content, and compaction by rolling.

Lime stabilization

In certain parts of Africa and North America, hydrated lime is used extensively for stabilization of soil, particularly for those with a high clay content. The hardening of soil-lime is accelerated at higher temperatures and the process is thus more suitable for use in warm climates. The strengthening effect of lime is much less than that of an equal quantity of Portland cement. The principal advantage of lime is the raising of the plastic limit of clayey soil, the soils becoming apparently drier, ensuring better pulverization and more uniform admixture of the stabilizing material. This quantity also makes lime suitable for the temporary stabilization of engineering sites to facilitate the movements of contractors' vehicles.

Quicklime is sometimes used to increase the 'drying' effect. A time interval between the first mixing in of the lime and the final pulverization assists in the drying out process, but this delay is not always necessary. Very rapid stabilization of water-logged sites has been achieved by quick-lime. The protection of the drivers of spreading and compacting equipment is essential.

The quantity of lime required to give the best results is usually from 4 to 6 per cent. Smaller proportions are sometimes used in pre-mixed work; larger percentages do not yield a proportionate increase in strength. Small additions of lime, 1 or 2 per cent, have been used in conjunction with cement for stabilizing soils which would be too heavy and wet for ordinary soil-cement. Lime is also used in conjunction with stabilization by bituminous materials.

Bitumen stabilization

The formation of sand-asphalts is another important method of soil stabilization. The technique employed varies with the prevailing climatic conditions, the choice of major methods depending on how rapidly the prepared soil dries out. Whatever method is employed, the object is to coat individual grains with bitumen, and to ensure this result the stabilizer may be used as an emulsion, as a cut-back bitumen, or as a cut-back bitumen mixed with wax or a wetting agent. Bitumen is incorporated in the sand to provide cohesion and waterproofing, and the types of soil generally treated in this way are clean sands or sands containing a small amount of cohesive material. In general, bituminous stabilization is of most value in granular soils. The stabilization of clays by bitumen, although possible, is un-economic, requiring the expenditure of a large amount of power.

Bituminous emulsions contain 40 to 50 per cent of water and carry the bitumen in a finely divided state. Further, this type of stabilizer is quite liquid and mixes well with sand, forming a slurry. As the water evaporates and the slurry dries out, the emulsion is broken and the bitumen deposited on the sand grains. When the moisture content drops to the optimum, compaction can be carried out. Bituminous emulsion is essentially a waterproofing agent preventing the rise of capillary moisture. The bearing value of the soil must be developed by suitable grading.

The success of stabilization by bituminous emulsion depends on the rapid drying out of the mixed material to a condition suitable for compaction. In Britain such drying out is problematical or slow, and other methods of using bitumen are of more value, although bituminous emulsions are assisted in drying out by the addition of a small quantity of Portland cement.

Sand-mix is the term describing the use of *cut-back* bitumen for the stabilization of sand. Bitumen is *cut-back* by the addition of a solvent or flux. When a light petroleum oil is used as a flux, the material is *rapid-curing* (RC). Oils such as certain paraffins give the MC or *medium-curing* grade, and a still heavier oil results in a *slow-curing* (SC) bitumen.

In Britain most of the sand to be stabilized is damp, there being few occasions when perfectly dry material is encountered. Ordinary cut-back bitumen will not mix easily with wet sand, and a special oil and special techniques have been developed for use under conditions encountered in this country.

In the *wet-sand mix* process the soil is first mixed with about 2 per cent of hydrated lime. Cut-back bitumen is next added, the cut-back being usually a proprietary material called Special Road Oil (SRO). The phenol incorporated in the cut-back forms a soap with the lime, causing a slight emulsification and so distributing the finely divided bitumen throughout the sand. There is also some reduction of surface tension and a consequent *wetting* effect. The smallest possible amount of binder is used, usually between 5 and 8 per cent. Wet-sand mix is best made with clean sand and the moisture content should normally be more than 4 per cent.

A type of cut-back bitumen incorporating paraffin wax, of which Shell Stabilizing Oil (SSO) is an example, is the most highly waterproofing of these stabilizers. The content of wax is normally about 4 per cent, but less petroleum binder is used (about 3 per cent) than in the wet-sand process, and the cohesion is provided by the clay content in the soil. As with other cut-back bitumens, 1 to 2 per cent of lime should first be incorporated in the soil. Waxed cut-back bitumen can be used cold or hot and gives a much better waterproofing effect than other bitumen mixes, maintaining the moisture content relatively constant.

Stabilization by resins

An important method of soil stabilization is by the use of resins. These are waterproofing agents having a similar action to the waxed cut-back bitumen. The soil should be pre-treated as for bitumen treatment by mixing it with about 1 per cent of lime. The resin used is a proprietary material and only about 1 to 2 per cent is required. Resins, like bitumen, extend the range of soils which may be stabilized to soils having poor grading and insufficient cohesion.

The resins, applied either as an intimately mixed powder or as a slurry, spread out in extremely thin water-resistant layers, within the voids of the soil, forming water-repellent barriers to capillary use. No other stabilizer is effective in such small quantities, 2 to 5 kg required per m^2 of 150 mm thickness. The material must, of course, be given optimum compaction.

Site operations

When the various tests on the available materials have been made and the type of stabilization decided upon, there remains the problem of translating these findings and decisions into practice. The two principal requirements are means of thoroughly mixing the materials into a homogeneous soil at the most suitable moisture content and means of adequate compaction. Whether for the natural subgrade or for artificially compounded base layers, the methods of compaction are those already described. The present section deals, therefore, wit mixing machinery and techniques.

Throughout the mixing operations and during all further steps in the production of the base, the longitudinal and transverse levels must be carefully checked and the surface level never permitted to deviate much from the required gradients. The first operation is a rough grading of the subgrade and the spreading over the surface of any material required to improve its mechanical stability. These operations can be performed by blade graders, which should have a blade about 3·5 m long. Graders may be tractor-drawn or self-propelled.

The soil is broken down so that the stabilizing agent may be thoroughly and intimately mixed into it. Pulverizing plant has been developed from the agricultural machinery originally used. The most effective work is done by rotary tillers after tractor-drawn ploughs have carried out the initial breaking-up. All the material should be well pulverized to a depth exceeding that of the compacted soil, and most of the soil should pass a 5 mm sieve. Large stones should be removed.

If stabilizing agents are to be added to a layer of soil, three methods of incorporation may be employed:

(i) *Plant-mix.* All the material passes through a stationary plant in which it is brought to the correct moisture content and in which the stabilizers are incorporated.

(ii) *Mix-in-place.* The whole of the mixing is done over the area of the site, the material being maintained at its correct level throughout.

(iii) *Travel-mix.* The material is piled into windrows, from which it is lifted by bucket elevator, passed through a mixing plant moving slowly along the windrow and discharged once more into a windrow of soil mixed ready for spreading and compaction.

Of the three methods, (ii) and (iii) are the most costly, (iii) being used only for large outputs and continuous working. Method (iii) is used for all types of stabilizer and (ii) chiefly for soil-cement and, sometimes, for resin stabilization.

Mix-in-place method for soil-cement. Five steps are identifiable in this process: pulverization, distribution of cement, dry mixing of cement, wet mixing, and compaction.

Distribution of the cement is best accomplished by a mechanical spreader which distributes the correct amount of Portland cement over the pulverized soil. The soil is then mixed by rotary tillers or pulverizers to a uniform colour and texture. This dry mixing is followed by grading and the addition of water to bring the moisture content to about 2 per cent above the optimum. The addition of the water and the wet mixing should be carried out as quickly as possible. The soil-cement should be cured in a damp condition, as is done for concrete. Single-pass machines, in which all the steps are carried out in one operation, are more efficient than multi-pass machines and can be used for a wider range of soils. They have several other advantages, such as that of affording the possibility of stopping the work on the approach of bad weather, or at a fixed time each evening.

Travel-mix method. The type of travel-mix plant usually employed consists of a loader and a mixer. The loader provides the motive power and comprises a bucket elevator which lifts the soil in the windrow into the pugmill mixer, where it is mixed with water and with liquid stabilizers when these are used. Powdered stabilizers, such as resin or Portland cement, are distributed ahead of the load and lifted from the windrown by the bucket elevator along with the soil.

After mixing, each windrow (whose size is sufficient to cover a 6 m wide strip) is spread and compacted. With Portland cement this should be done as quickly as possible before setting of the cement takes place, but with bitumen the final spreading may be delayed if found necessary.

The windrow and travel-mix method of operation is very much less liable to disturbance by weather conditions than is the mix-in-place method. There is little or no evaporation from the interior of the windrow, and rain runs off the steep sides without seriously altering the moisture content.

11.4 REINFORCEMENT OF GROUND

There is now an imperative necessity to build on sites which, only one or two decades ago, would have been considered quite unsuitable to take the load of a structure. Power stations and factories requiring large supplies of water must be founded near estuaries, sites must be built on again after being derelict or used as convenient dumping grounds. As a result, there has been a growth in experiment and experience in the stabilizing of areas in weak alluvium, in industrial fill, or in other soft material. The methods of soil stabilization suitable for the relatively shallow depths adequate for road construction cannot be effectively applied to building foundations.

Normally, if a building must be founded on a site where the upper strata are weak, it is supported by piles taken down to a firm stratum (if one can be found) or the net load is reduced by the use of buoyant foundations. Either of these solutions is costly, the second being particularly difficult in very soft ground. Improvement of the surface layers down to a depth of several metres so that the building may be founded nearer the existing ground level is an alternative which has been pursued by several techniques. The concomitant problem of retaining soil so that it presents a vertical or nearly vertical face (for example, for roadways round buildings) is also one receiving increasing attention. A less costly solution is required than that offered by the conventional reinforced concrete retaining wall.

The principal methods are described below, but none can be looked upon as having reached its final development. The present state of the work shows how it is possible to apply basic principles to be of direct value in supporting foundation loads on poor ground. In the technique known as *dynamic consolidation* or *deep compaction,* the soil itself is improved in its supporting power by the application of mechanical tamping with energy inputs of unusually large values. The second method is also a compaction method – *vibro-compaction* – but the compaction is applied not to the existing soil, but to granular material added over specific areas to form stone columns within the weak foundation soil. The third technique discussed is that of reinforcing the soil, not to carry imposed loads, but to carry its own weight (and some superload, perhaps) without slipping when a vertical or nearly vertical face is formed. The soil is not supported by an additional structure such as a retaining wall, but is assisted by reinforcement to be self-supporting. This is usually known as *reinforced earth* or *terre armée.*

Deep compaction
The present-day necessity of building on fill, on alluvium and other soils normally unsuited to the support of conventional foundations

has led to several methods of improvement of the soil for the support of buildings. The technique under review in this section is a means of compaction by the dropping of heavy tamping loads from considerable heights on to the poor material requiring improvement. From 10 to 20 tonne can be dropped through heights of up to 20 m or more. Such heavy tamping, if done on granular cohesionless soil, causes compaction by driving out air, and this compaction occurs through a considerable depth from the surface (10 to 15 m). In such soils, the success of compaction by the application of high energies is assured.

Unsuitable soils, however, are more likely to contain silt and clay particles, and to have low values of permeability. By experiment it has been found that the technique, applied to saturated cohesive soils, has the effect of producing very high pore pressures which, in turn, produce a partial liquefaction of the soil. The increased void ratio resulting from the break-up of the soil structure, produces a higher permeability, and so allows of a rapid dissipation of pore pressure. Thus, although this technique consists of the application of high-energy tamping, it produces consolidation of cohesive soils faster than would be possible under static loading even with the assistance of sand drains.

The effect of this on settlement can be expected to be striking. Although pre-settlement of the soil can be achieved by loading the surface heavily and statically, such preloading under the best conditions gives somewhat less consolidation and settlement than that obtained under the finished construction. With dynamic consolidation, on the other hand, it is claimed that two to three times the ultimate settlement under the constructional load is achieved by pre-treatment by high energy tamping. This process, because of the heavy blows used and the vibrations produced, has its field of application somewhat restricted to wide isolated areas remote from buildings. It is, however, in such areas that weak alluvium and industrial fill occur. On these materials dynamic consolidation shows to good advantage.

Vibrated stone columns or vibro-compaction
The practice of making holes in soft material and back-filling with crushed stone to form a firmer foundation is not modern; it has a history of well over a century. But with the advent of powerful

vibrators, first used for placing concrete, the technique developed into a now well known feature of foundation engineering. It is used for founding buildings and even embankments on poor material such as random industrial fill, estuarine deposits and other soft sediments. Settlement is much reduced if compared with settlement on the original ground.

The present day method of penetrating the weak ground consists of using an enclosed tubular vibrator of about 0·5 m in diameter, with a length to suit the job required – holes range from 5 to 10 m, in normal average conditions. The tube is vibrated at high frequency and lowered into the soft ground, assisted by jetting with air or water. Air is used for the harder materials and water for the softer. The two techniques (dry and wet) are called *vibro-replacement* and *vibro-flotation*, depending on whether the jetting is done by air or by water. When the vibrating tube has penetrated to the required depth, which happens quite rapidly, gravel is provided and the tube withdrawn, compaction being carried out by a reciprocating movement of the tube, as the stone is filled in. Each hole, as measured by the volume of the tube, may be only 1 or 2 m^3 but it can absorb much more volume, the stone being forced laterally into the soft, peaty or random material, or loose sand strata. A figure giving the order of the absorption of stone, is about $\frac{1}{2}$ m^3 m^{-1} of length of column.

The stone columns, so formed, may carry a concentrated load from a pad footing, or strip loadings when the columns are in rows, or they may be distributed to a triangular pattern across the site. Suitable spacings vary from 1·5 to over 2·5 m between adjacent columns.

Confirmatory published researches on the bearing capacity of these columns are not numerous, although two authors, at least, have reached similar conclusions. A sound preliminary estimate to be checked by loading tests, if possible, is that the allowable bearing pressure (taking into effect the possible settlement) is about 8 to $8\frac{1}{2}$ times the undrained cohesive strength of the soil. It is in cohesive soils that there is the greatest need for support. The support afforded by the stone column to the structure above depends on the friction between the stone and the soil. It should be remembered that the lateral penetration of the stone into the surrounding soft material by reason of the compacting and vibrating effect of the *vibroflot* – the vibrating tube – gives extra support by providing an accentuated frictional effect. Clay mixed with random industrial fill is readily strengthened by this technique.

Ground anchors and reinforced earth

In an earlier chapter the earth-retaining technique of *anchored sheet-pile walls* is discussed. The ties used in these walls act in support of the sheet piling, against the earth pressure by calling on the passive resistance exerted by the anchors to the ties. No reliance is placed on the friction between the anchor shaft and the soil.

In Chapter 10 *ground anchors* show this principle of the use of tension to support earth masses developed still further. Here, the support is almost entirely by friction over the operative part of the anchor, with some extra effect from the under-reamed portions of the anchor shaft.

When 'almost entirely by friction' becomes 'entirely by friction' and the number of supporting units becomes more numerous even than that in a ground anchor system, we are in the realm of *reinforced earth*. The object of this system developed by Henri Vidal is to make the soil and the reinforcing units act as one body. The reinforcement consists of metal strips laid close together in layers, and stretching at right angles to the face of the earth bank to be supported. Each layer is separated from those above and below by thin layers of soil. There is a system of interlocking facing units attached to the ends of the reinforcing strips, and forming a vertical face to the self-supporting reinforced earth mass.

Diaphragm walls

This method of giving ground support has become a regular feature of sub-surface construction work. It has advantages over similar methods of providing a gel or of freezing the ground. It is particularly used for the construction of reinforced concrete walls or diaphragms, below the general ground level, before any excavation is made. The technique is to make the excavation for the wall (for basements, cut-offs, etc.) and to fill this excavation with a bentonite suspension. Bentonite is a clay-like sediment which originated from wind-blown volcanic ash. Its composition is largely colloidal silica, which is capable of absorbing large quantities of water.

Colloidal suspensions of this kind are thixotropic; they form gels which break down under the input of energy and gel again when left quiescent. This property is put to good effect when the wall excavation is filled with bentonite suspension, for this first penetrates to all the interstices of the surface of the excavation and then gels on being left to settle. This gel applies a supporting pressure on the sides of the excavation.

Reinforcing cages are lowered into the bentonite suspension, and concrete placed by tremie pipe from the bottom upwards as is usual in underwater concreting. The concrete is not contaminated by the bentonite, nor does the bentonite adhere to the reinforcement; a good bond is achieved. This I.C.O.S. technique is of Italian origin (Dr Christian Veder) and is, for some specialized work, an excellent means of temporarily reinforcing the soil, particularly in urban areas where disturbance and noise must be at a minimum. The method can also be used effectively with post stressing of the main vertical bars.

CHAPTER 12

Site investigation

The procedures in *site investigation* may vary from the visual examination of soil taken from small test pits to an elaborate soil survey involving numerous deep boreholes. The Code of Practice 'Site Investigations' summarizes the information which it is desirable to obtain, and describes the methods used for procuring it. In the context of this chapter, topographical surveys and levelling surveys are not included in the term 'site investigation' although they are of great importance in the total study of the site.

12.1 GEOTECHNICAL INVESTIGATION

The term *site investigation* covers at least two broad stages of inspection and testing. The *first stage* is that of establishing and recording the state of the soil on the selected site. This must be carried out to a depth below which the new construction will be unlikely to effect any significant change, especially in conditions of stress in the soil. Concurrent with the study of the soil itself is the parallel investigation of ground-water conditions. This is of great importance and is sometimes taken sufficiently seriously.

The *second stage* is one which has become of increasing value. It concerns the study of the principal changes which take place after a new construction has been completed, or even which occur during its erection. These changes take place in the soil whose original state has been determined during the first stage of the investigation. Sometimes the eventual state of the soil after continuous loading is of more value and interest than the originally measured conditions. The effect of the interaction of soil and structure is also evident in the structure itself – the settlement of a b building or the changes in pore-pressure which take place within a structure such as an earth dam. The object of this second stage is to correlate theoretical preditions with what occurs in practice. The findings are of value in later predictions. Needless to say, since continuous record-keeping is costly and can have but a small effect on a structure alread erected, it is not always pursued. Large constructions, however, are often instrumented and the results published as research reports to be used as sources of information in similar circumstances in the future.

There are three principal uses for the techniques employed in the first and second stages of site investigation. These phases are:

(i) the assessment of the properties of extracted or quarried material, with a view to deciding on their suitability for engineering construction (concrete aggregates and sands; material to be used for fill or for the construction of embankments);

(ii) the study of existing conditions in the sites for foundations of new works;

(iii) the assessment of the stability of proposed structures in advance of construction.

Phase (iii) also includes decisions on the causes of failure of existing works (although this should not be looked upon as a normal operation!). It is in phase (iii) that the findings of the *second stage* of site investigation, mentioned above, can be put to their most valuable uses.

Investigation of sites for new works
Sites may be subdivided into:

(i) *compact* or *area* sites, such as sites for buildings, bridges, or dams;

(ii) *extended* or *linear* sites, such as routes for railways, roads, or canals.

In the first type of site the area of ground to be investigated is usually well defined, though sometimes it may be possible to examine alternative sites. In the extended type of site investigation, such as for a road or railway, the soil conditions on alternative locations may have a considerable influence on the final selection of the route.

The extent of the investigation which should be undertaken for a particular project depends on a number of factors, including:

(i) the size, type, and importance of the proposed works;
(ii) information already available regarding the nature and variability of the soil;
(iii) time available for the investigation.

For example, if the soil strata are already known to be good from the engineering point of view, and fairly uniform over a large area and to a considerable depth, a minimum of site investigation will be necessary. On the other hand, variable strata and the presence of 'difficult' soils will necessitate a more detailed survey. Soils which are likely to cause trouble include peat and soft clays (low shear strength and high compressibility), silts (low strength and susceptibility to frost action), and fine sands (liability to exhibit quicksand properties). The cost of the investigation must be weighed against the probable saving in the cost of the foundation, but it should be borne in mind that even lavish expenditure on foundations to make up for ignorance of the soil properties will not ensure safety if an unsuitable type of design has been adopted. The decision as to the extent and nature of the site investigation necessary must depend to a very large extent upon the judgement and experience of the engineer.

The information to be sought for in any site exploration includes the following:

(i) the nature, thickness, and dip of all soil strata in the region likely to be affected by the proposed work;
(ii) the properties of the various soils occurring in this region, e.g. the density, natural moisture content, shear strength and compressibility;
(iii) the ground-water level, its flow and, if possible, seasonal variations.

In order to obtain this information, an adequate number of samples of soil must be taken for identification and for field and laboratory testing.

As a guide to the depth to which the exploration should be carried it should be remembered that the pressure bulb for one fifth of the contact pressure imposed by a circular loaded area has a depth of 1·5 times the diameter of the loaded area and that adjacent loaded areas have a cumulative effect. Sometimes the existence of solid rock at a depth less than this will obviate the necessity of deeper exploration. On the other hand, the presence of soft soil at this level may necessitate penetrating to a greater depth. For example, in exploring a compressible stratum it is desirable to ascertain its total thickness and whether or not still lower strata are permeable. Other factors which should be taken into consideration in deciding upon the depth of exploration are the seasonal limits of ground-water level and the known or suspected presence of mines or other underground workings.

Investigation for stability

The object of investigating the properties of the soil at any site is to determine the stability of the structure concerned, and to ensure an economic design. This is true whether the structure is extraneous to the soil, or is itself a soil structure such as a dam, an embankment, or a cutting. The samples of material and their properties should lead to a determination of the bearing capacity of the natural soil and its possible deformation under the proposed load. This necessitates not only the measurement of the properties of extracted samples by testing in the laboratory, but also the determination of data from *in situ* measurements.

12.2 THE DESK STUDY

The engineer in charge of the site investigation must always be briefed on the exact reasons which prompted the request for a site investigation. A great deal of extra expense in boring and testing can be avoided if the scope and intensity of the investigation can be adapted to the client's specific requirements much more closely than sometimes occurs. With this knowledge in mind, the engineer should then walk over the site, observing not only the site area, but also the surrounding features. With experience, this can be a rewarding exercise leading to an intelligent and productive use of what is commonly called a *desk study*.

The elements listed in this section belong to the *desk study,* but other investigations may well be required for the particular circumstances of the site being studied. References are made to British sources of information; similar records are kept in other countries, but in more remote and primitive areas the investigator may be thrown much more on his own resources without the assistance of long-term records.

Geological maps

Since the work is largely concerned with geology, knowledge of the geological formation of the site itself and of the surrounding area is of first importance. In the United Kingdom very adequate maps are published, and the Institute of Geological Sciences and its regional offices can provide further information for particular areas. Both *drift* and *solid* geology may require study if the structure is large. Even in small structures, the possibility of the presence of base rock near the surface is of interest and importance. Technical papers published about the area within which the site lies may give information not available on any of the geological maps. *Handbooks* are also available on the regional geology of the U.K., and similar maps and publications available for other countries. When the survey of geology is completed, it is sometimes advisable to consult the *Soil Survey of Great Britain* (Agricultual Research Council) and the *Land Utilization Survey* (King's College, Strand, London and the National Library of Scotland, Edinburgh). Both give details of the kinds of soils to be expected. The engineer/geologist should also make his own maps showing visible features such as natural surface drainage patterns, angles of slope of the surface, and identifiable geological forms (scarps, drumlins, etc.).

Mining records

In coal-mining areas, the presence of underground operations and the dates at which these operations have taken place or will take place must be investigated if the site is over workings. In the U.K., the National Coal Board, through its regional offices, can provide detailed information about present and future conditions on specified sites. Salt mining also offers possibility of subsidence at the surface. For these and any other type of mining operations it is essential to include underground conditions in the desk study. Often a small movement of the proposed siting of a building may protect it from excessive differential settlement in the future.

Services

The use of sites which have already carried structures of some kind is now commonplace. Extension of existing buildings may also have to be constructed on already-used ground. All sites should be examined, and their history studied to determine the possible presence of old walls, abandoned machinery and waste material. On one site, a completely buried basement was found to be solidly packed with old gas meters! Principally, this aspect of the desk study should be concerned with the locating of service cables, pipes and ducts. Service authorities now usually have adequate records, but it is still common to have to guess the location of services laid in the days before exact recording. Much damage to service supplies occurs accidentally each year, during site works. It is part of the responsibility of the site investigator to ensure that, as far as is possible, the locations of service ducts and cables are known.

Meteorological records

Moisture plays a very important part in all soil problems. It is, therefore, essential to collect as much preliminary information as possible on climatic conditions, especially in countries where wide variations occur. Data should be sought on the amount and seasonal variation of all kinds of precipitation, particularly with reference to maximum intensities and the possibility of the flooding of the site. In the U.K., detailed records of long-term observations are maintained. Information can be obtained from the Meteorological Office and local records are also available. Such data are often of importance in investigations both before and after the construction of buildings or civil engineering works.

Ground water

The level of water in the ground – the *water table* – will be recorded in the boreholes during the site borings. This refers only to the period of the investigation, but a long-term study of the fluctuations of the water table is required if full knowledge of the problems of the site is to be obtained. It is advisable to leave standpipes in some of the boreholes as long as possible, but even this technique will not provide all the information needed as the time involved is much too short, especially if the site investigation is called for towards the end of the planning period. The flows in nearby rivers can indicate some of the possible variations in water level of the district, and these can be related to the site if it is not far from the river. *The Surface Water Year Book* gives records of flows for a number of gauging stations, and records from other stations may be obtained on request. On the coast, where construction can be close to or in the sea or harbour areas, the charts prepared by the Admiralty must be consulted. They provide valuable information on sea levels and soundings.

Early maps
The early maps of a city can often provide information of importance for a modern site. They can supplement the information obtained on mining records and services. They may also show former topographical details of the site area. The former existence of buildings on the site are shown, and even the underground conditions can be explained by early maps. In the investigation of the site for a large building, one borehole showed a large depth of silt which had not appeared in the other borings. Old maps showed the presence of a seventeenth-century mill pond of considerable size, and appropriate measures were able to be taken in the design of the building foundations. On another site, the depth to bedrock was shown as about 2 m in one borehole, and in adjacent one, 15 m. A map of a century before showed the presence of a small quarry, although later maps showed nothing, and there was no evidence in the appearance of the ground. This small extraction of stone for a local purpose might have caused great expense if the extent of the buried quarry had not been determined before foundations were designed.

Air photography
The principal use of air photography is in map-making, but it has also been extensively used as a means of identification of natural and artificial features.

For classification of soils from air photographs, some knowledge of physical geography and geology is necessary. Further, more, experience in the interpretation of the tones of the photographic image must be gained. The tone of a soil formation depends on the direction of the incident light and also varies with the texture and moisture content of the material, and when the surface of the ground has been disturbed its texture will remain different from that of the surrounding land. This has been a vital factor in archaelogical exploration from the air. Tracks and foundation lines, long since rendered invisible on the ground, show up clearly on air photographs.

A ready means is thus available for the detection of areas of made-up or disturbed ground as a first step towards more localized and detailed inspection.

Broad geological features are easily identified on air photographs. The photo-geologist determines the directions of faults and dykes and the presence of mineral veins, and by skilled deduction and inference from the photo-cover of a whole area may extract much additional information.

When the precise identification of soil types is in question, caution must be exercised. The surest method of general stratigraphical classification is by the study of vegetation and land form. The latter generally indicates the age of the formation, and the type of soil resulting from the disintegration of the geological structure is closely related to the vegetation. *Terrain evaluation* from aerial photographs is now a recognized technique which has been developed to a high standard of accuracy. It has been used by consulting engineers in locating routes for roads in undeveloped countries.

The smallest scale of air photography which allows of broad interpretation of soil conditions is 1/20 000, but the information to be extracted at this scale is limited unless the photographs are viewed stereoscopically. Subsequently, for detailed examination of particular sites, a larger scale of photograph is required, and such photographs should be used in conjuction with direct inspection of the ground.

12.3 TYPES OF SAMPLE

The object of obtaining samples of soil at various horizons in a soil mass is to enable classification to be carried out, and the engineering characteristics to be determined by testing. Sometimes a shallow pit gives sufficient access to the strata concerned, but more frequently samples must be obtained from boreholes at depths beyond the reach of shallow pits. Methods have been devised and are continually being improved to obtain samples in what is known as the *undisturbed* condition, although it is obvious that disturbance must take place during extraction. The term is used to describe a sample in which the soil structure has not been greatly altered, and there has been no change in moisture content, void ratio and chemical composition. From samples of this kind reliable estimations can be made of the probable behaviour of the soil under load.

In addition to the *undisturbed samples*, which are the most important, there are disturbed samples from which certain characteristics of the soil can be determined. These take the form of

(i) *representative* or *jar* samples;
(ii) *bulk* samples.

The three types of sample are each used to extract information about the strata according to the following broad pattern:

Representative or jar samples. Inspection of these samples in the laboratory allows the engineer to describe the physical state and characteristics of the soil. Such descriptions should never be left to the driller on the site. He will give a description of each stratum briefly, in order to maintain a distinction between samples, but the laboratory examination is still necessary if the soil is to be described accurately. Description of soils as *soft, firm, stiff, very stiff* are to some extent subjective, but are well understood. Clays must be described according to the inclusions, such as *partings of fine sand,* or *laminations* or *inclusions of gravel.* Granular soils should be described with some reference to sizes of particles and to grading. Representative samples can also be used to determine moisture content and secondary classification characteristics, but these are more frequently determined from the undisturbed samples, or bulk samples.

Bulk samples. These are also disturbed, and are merely larger representative samples. They are used for *classification tests* such as *mechanical analyses, consistency limits, specific gravity, moisture content, chemical analyses, compaction characteristics* and *density.* The results of these are instructive to the engineer in charge of the design of foundations or of soil structures. For both jar and bulk samples it is important that they should be protected against the loss of moisture by evaporation and that they should be used for testing very soon after being extracted.

Undisturbed samples. These are the most valuable samples, as they are used for obtaining the *cohesive* and *shearing strength* of soils and, from specially cut samples, the *consolidation* characteristics, and the value of *permeability.* Disturbance, in the sense described above, can take place not only during the extraction process, but also after the sample has been obtained. If the sample is not adequately sealed from the air, there is a danger of a change in moisture content and even of oxidation. If the sample is subjected to freezing temperatures or damaged in extraction from its container, the tests on soil properties may be misleading. For these reasons, and to avoid the false characteristics resulting from the inclusion of larger stones in cohesive materials, undisturbed samples are being extracted and tested in larger sizes. This tendency towards increasing the dimensions of undisturbed samples assists in the more accurate assessment of soils for their engineering characteristics.

Ground-water samples. Routine extraction and testing of samples of ground-water is very important, and must not be forgotten. Chemicals, particularly sulphates, may exist in the ground to such an extent as to be deleterious to concrete foundations.

12.4 SAMPLING AND TESTING

The extraction of bulk samples or of representative samples in screw-topped jars is relatively simple compared with the difficulties encountered in extracting samples with so little disturbance that they are substantially in the state in which they existed *in situ*. Sampling is carried out either in boreholes or in pits accessible from ground level.

Sampling from boreholes

Samples taken from boreholes if in the 'undisturbed' state are usually of cohesive soil. Samples 100 mm in diameter can be tested at the full diameter, or can have smaller diameter samples (38 mm) extracted from them. These $U(100)$ samples, as they are called, are the normal for triaxial testing. Sampling tubes are fitted with a cutting edge, the inner diameter of which is slightly less than the inner diameter of the tube. This allows for slight expansion of the soil as it enters the tube. A smooth, well-greased inner surface also helps to reduce the friction and so minimize disturbance of the structure of the soil. Similarly, to facilitate driving, the outer diameter of the cutting nose is slightly greater than that of the body of the tube, but the difference must not be so great as to cause difficulty in withdrawing the tube. A non-return valve at the top allows air to escape as the sampler is being forced into the soil, and helps to retain the core as the sampler is being raised. In this design the cutting nose and the driving head are detachable. The body of the sampler can then be used as a container for the sample until it reaches the laboratory. The samples are packed as described below for transmission to the laboratory.

The Area Ratio of a sampling tube is the difference of the external and internal cross-sectional areas, expressed as a percentage of the internal area. The figure should be as small as possible consistent with strength. The thinner the tube, the smaller the area ratio, which is usually about 10 to 12 per cent.

Even cohesive samples sometimes fail to remain in the sampling tube as it is withdrawn up the borehole, if the soil is very wet. To prevent this loss of sample, *core catchers* have been devised, which trap and hold the core which has been cut. They are of two types, the spring type and the hinge type, nd both make it less likely that samples should be lost.

The extraction of relatively undisturbed samples of granular soil from boreholes, is difficult, but devices have been developed to retrieve samples from which, it is claimed, the undisturbed properties can be measured with fair accuracy. It is more usual, however, to attempt to determine the properties of granular soils *in situ* and by means of one of the *penetration tests*.

The development of *sea bed sampling* for use in exploration for oil deposits has resulted in better coring and extraction of sediments of all kinds, with samples from about 60 to over 100 mm and core lengths of up to 6 m. The Institute of Geological Sciences (Scottish Continental Shelf Unit) has collaborated with manufacturers to ensure maximum recovery of sample with core catchers in the cutter head.

Piston samplers have been devised in which an outer cyliner and an inner piston tube with a penetration head are both driven together. When the depth is reached at which sampling is required, the piston is withdrawn upwards into the top of the tube and the tube is then driven further to take the sample. For extraction, a vacuum can be maintained by the piston head above the enclosed sample, and a full recovery of sample is possible.

Drilling techniques are used in order to retrieve samples, and it is of great importance that these samples do accurately represent the soil layers passes through, both in composition and in thickness. The young engineer must appreciate the necessity for regular and informed supervision of the extraction and recording of samples. *Site supervision* must take a high place in the planning and costing of any site investigation if the conclusions are to be reliable. A beautifully prepared boring record is of no value at all unless some experienced engineer has made sure that it truly represents the conditions below the surface.

Shallow sampling

Test pits can be used in various ways. They give the engineer the opportunity of seeing the upper strata in bulk instead of merely in small samples taken from a borehole. They can confirm conclusions reached from the more extensive a detailed borehole sampling. It is not often that trial pits can be relied upon as the only source of information, even if they are taken down to foundation level. The stability of a foundation is much more dependent on what lies below the bottom of a trial pit than on what appears within its depth.

Disturbed samples can easily be dug from the sides or bottom of a trial pit. Undisturbed samples of some size can also be carefully extracted after being excavated carefully with a spade. A core cutter, however, is more effective for clay and can produce a sufficiently large undisturbed sample. Small samples can be extracted by driving in a 38 mm sampling tube and digging it out carefully. Such samples can be checked by unconfined compression on the site.

For non-cohesive soils without a high gravel content, a sampling box or tube, open at both ends, can be driven or pushed in to its full depth and the surrounding soil removed. A trowel is introduced under the box and the whole is reversed. The bottom is now levelled off and a bottom cover put on.

A modification of this method can be used for cohesive soils which contain gravel (where it is difficult to push in a sampling cutter). The soil is dug away to leave a column of undisturbed soil standing in the middle of the excavation. It should be slightly smaller in dimensions than the sampling box to be used. The box is placed over the sample, and the spaces between sides and sample packed with damp sand. *In situ* tests with a large shear box have been made on such undisturbed pillars of soil at the bottom of a trial pit.

Handling of samples

The undisturbed samples or cores must be adequately protected against change of moisture content and damage in transit. When the sample is to be retained in the tube, which is the usual procedure, it is necessary to cover the exposed ends of the sample to avoid loss of moisture. Tightly screwed caps are used, and a layer of paraffin wax is useful. Other devices have been invented such as a plug containing a rubber membrane which extrudes round the circumference when the plug is tightened, and thus effectively seals any possible peripheral leaks of moisture.

The representative samples are normally contained in jars with screwed lids, or in similar plastic containers. The much larger bulk sample is usually taken to the laboratory in a plastic bag. It must be remembered that these samples, however well contained, dry out and become useless to the investigator. If, however, the samples are examined, described and tested immediately after they are received in the laboratory, these methods of dispatch are acceptable.

A careful record should be kept of all samples, together with a written description. All the data regarding location and type of material appears in the boring record and in the record of laboratory testing. The *U(100)* tubes should be identified by labels in such a way as to make it quite certain that no mistake in identification can be made. The top and bottom of the sample should be separately identified.

Extent of testing necessary
There is sometimes a tendency amongst architects and engineers, not familiar with what a site investigation is supposed to accomplish, to order every test possible on all samples. This is uneconomic and unnecessary. The testing programme and the properties of the soil which should be determined vary considerably with the circumstances of the site and of the structure to be erected. For example, if the investigation is in preparation for the design of a building foundation, the exploration and testing can vary from a few shallow trial holes in stiff boulder clay for visual inspection, with check tests in unconfined compression, to several deep boreholes with an extensive programme of *in situ* and *laboratory* testing.

Bearing capacity is not a single property of a soil; its value varies, not only with differences in the soil, but also with the depth of the foundation and with the nature and shape of the substructure. The best approach is to make a preliminary judgement on the type of substructure and to base the scale of the investigation on that. For a lightly loaded warehouse framework on widely-spaced pad footings at or close to the surface elaborate testing is likely to be unnecessary, and some small settlement would not be considered damaging. On the other hand, for a multi-storey building where soft strata are expected, the testing would be much more comprehensive in order to determine the properties and coefficients from which a full understanding of complex soil conditions can be obtained. *In situ* and *laboratory testing* are considered in the next two chapters.

Range of tests required during the site investigation report and after it is completed

Three categories of tests are given below, arranged in groups according to the use to be made of the results. In each category, a range of tests is given. For any particular site or soil mechanics problem, a selection of these tests must be made. Other information such as unusual geological conditions or the presence of mining activities may also have to be added to the date accumulated for any site. No explicit distinction is made at this point between *in situ* and *laboratory* tests.

Design data. Results of tests from this range are required before an effective design of soil structures or foundations can be made (in alphabetical order).

Borehole records; California bearing ratio; chemical analyses; classification (primary and secondary); clay minerals (types present); cohesive strength; compaction characteristics; consolidation characteristics; densities (in situ and relative); freezing (resistance to); model tests; moisture contents; penetration resistance; permeabilities; pore-pressures; shearing resistance (angle of); specific gravity; surveys (geophysical and topographical with particular attention to borehole or trial-pit levels of the surface and strata, related to ordnance datum); void ratios; water tables (and fluctuation of).

Control data. While construction is in progress, it is important on the site to accept or reject materials, to check the performance of contractors against the specification, and to bring to light and measure any unsuspected variations in conditions which might lead to a change in the design. Some of the tests likely to be required are shown below. Most can be carried out in a site laboratory, but some may require attention at a central and more fully equipped laboratory.

Cohesive strength; densities (in situ); compaction characteristics; classification (primary and secondary); moisture contents; pore-pressure; shearing resistance (angles of); water tabel levels; trial constructions.

Research data. In soil mechanics there are wide variations in conditions; data are extracted from heterogeneous deposits and loadings cannot always be defined exactly. Any check on how closely the behaviour of a structure follows the predicted pattern is of value. The tests required in the monitoring (sometimes over a long period) of the pressures and movements occurring in structures are usually demanding. The apparatus is more specialized, delicate or complex. The results obtained from many sites are coalesced into bodies of knowledge to guide later designers. The range of these monitoring tests is under constant review, but covers such subjects as:

Contact stresses between foundation and soil; total stresses with a soil mass; earth pressures exerted on new construction; behaviour of underground structures such as tunnels or deep basements; displacement of ground under or near new buildings; loading tests on foundations or piles; permeability measured in situ; pore-pressures and their variations; monitoring of seepage; displacement measured by inclinometers; settlement of structures monitored over long periods; vibrations transmitted through soils.

12.5 ORGANIZING EFFECTIVE SITE INVESTIGATIONS

Site investigations can be divided into three stages:

(A) Boring, drilling, the extraction of samples and the carrying out of site tests;
(B) Preparation of the samples extracted and their testing and examination in the laboratory;
(C) Assessment of the meaning of the facts extracted and the writing of the Site Investigation Report.

Since the operation is directed to discovering unknown facts about what lies hidden below the surface of the ground, exact predictions of the extent of the investigation or of its cost cannot be precisely made although, from past experience, it is usually possible to give a rough approximation to a 'budget' cost. Many contracting firms, however, are willing to tender for all three stages (A), (B) and (C) and *within the quoted price* to give facts and predictions.

When difficult conditions are encountered, this simple contractural method is not adapted to producing the best results. Discussion amongst those professionally concerned, has shown that there is a general opinion that stages (A), (B) and (C) should be considered separately, even if all are carried out by the same organization.

Schedule of rates for boring, drilling, sampling, *in situ* testing and laboratory testing allow of a flexible control. Stage (A) is best and most economically handled under the usual contractural procedure, and stages (B) and (C) carried out by specialist consultants.

There are several operational points which must be kept in mind if the investigation is to be economically prosecuted and is to give the information required for the particular construction proposed.

(i) Boring, and the consideration of the foundation materials encountered should not be restricted closely within the limits of the construction proposed. Often, a wider investigation shows valuable data on, for example, the dip of strata, variations in the water table, and the extent and drainage of strata underlying the building concerned.

(ii) The depths to which exploration should proceed should be estimated ahead of the work, but it should always be possible to give immediate decision on the stopping of the boring, or on its extension to deeper levels if these steps help the efficiency of the work.

(iii) Supervision of the work on site must be close, and carried out by qualified operators or by engineers. Means must be provided for quick decisions to be taken by the engineer in charge of the work if an alteration in the boring or site testing programme is required.

(iv) Many different tests may be made on site, but a multiplicity of these, or the carrying out of all of them does not make the investigation more exact. The types of test to be made are dictated by the purpose for which the site investigation is required. Sometimes an elaborate programme is necessary; sometimes a few simple tests may suffice. An effective site investigation reports only those tests required for the solution of the problem in hand.

(v) The amount of work in the writing of the report – which entails the interpretation of complex and interrelated facts – cannot be predicted exactly. The interpretation of results and the preparation of the report should be valued on an hourly scale. Sometimes the engineer in charge of the design requires only a factual statement concerning the findings, and prefers to interpret them himself, making the necessary decisions concerning foundations, without external advice. For both types of reporting, the facts presented must refer, in position and level, to precise locations on and below the surface. Levels should always be referred to the statutory datum (Ordnance datum in U.K.).

(vi) The main investigation, if it is likely to be large, should be preceded by a preliminary investigation by a few boreholes. The engineers can then plan the main investigation to provide the greatest possible body of information within the limits of the problem in question. The best programme is that which allows of feedback of information from preliminary studies to the later site investigation, and from all stages of the site investigation, continuously to the engineering design team. If the engineering design must await the final production of the bound and glossy report, the latter may come too late. The site investigations in all their stages do not form achievements in their own right, but must act as supports to effective design throughout the design period. For this reason, site investigations should be commissioned well in advance of even the preliminary design stage.

The site investigation report
The principal object of making studies and tests of soils, apart from their intrinsic value as experimental exercises, is to allow the engineer to predict the behaviour of foundations or of structures composed of soil or supporting a soil mass. The structures erected on foundations designed as a result of studies of soil must carry their loads effectively and without showing signs of movement. The retaining walls, cuttings, embankments and natural slopes, studied with a view to assuring their stability, should demonstrate that stability after having been adapted physically to resist the imposed conditions. A second objective of studying and reporting on soils is that of assuring that structures of all types, whether consisting of soil masses or supporting artificial loads on soil foundations, should provide their services at an economical cost. These objectives of *stability* and *economy* can be achieved efficiently only when the results of the site investigation are properly integrated into the planning and design processes.

The study of the site in three dimensions and the results of the soil tests made come into focus in the *Site-investigation Report*. To this focus also come the requirements of the proposed structure. The pages of a comprehensive Site Investigation Report form the final presentation and discussion of how structural design may be matched to the properties of the soil. The Report indicates the planning required to balance the structural needs on one hand with the strength or vulnerability of the soil on the other. The Site Investigation Report can be looked upon as the end product of the practical study of soils.

For effective use of a site investigation, there must be a link between the requirements of the structure proposed and the studies carried out on the soil. In preparation for civil or structural engineering construction there cannot be a totally unqualified site investigation. The investigation and report must be directed to giving advice (and warnings if necessary) *in relation to a particular form and extent of construction.* It is not sufficiently precise for the client to ask for 'a site investigation' without qualification, merely because the site is to be used for some form of construction. It is essential, if money and time are not to be wasted, for a brief to be prepared of the type (or alternative types) of construction proposed, their proposed location on the site, the Ordnance datum level at which the structure is to be usable (whether, for example, basements are required), whether there is to be, perhaps, pumping for a deep well for water supply, and other conditions the client has in mind. The site investigation for one form of construction is different from that required for another. The types of test and the range of testing are all influenced by structural requirements.

If this linking of requirements and investigation is to be fruitful, the need for collaboration between the client and the site investigation team immediately becomes apparent. The site investigation must be undertaken at a stage in the preparation which allows of it becoming part of the planning and design process. When working drawings are beginning to take shape on the drawing boards, the appointment of a site investigation team is too late in the programme. On the other hand, a complete site investigation carried out when the proposals are in no more than a nebulous state may well contain whole sections which, in the end, will be found to be unnecessary.

It is, by no means, easy to time the site investigation so that:

(i) adequate time is available for its completion and report;
(ii) it is not carried out so late that it can only confirm design decisions already taken, or cast doubt on such decisions too late for alterations to be made;
(iii) it is not carried out so early that the results cannot be related to the detailed design requirements of construction, and can thus have only a minor effect on planning.

The objective of using the Site Investigation Report as a defined influence on structural decisions must be kept in mind. If the soil conditions on site are properly integrated with the structural proposals, the final result will be much more effective and economical. The cost of a full site investigation is a minor part of the cost of construction.

The Site Investigation Report must, therefore, be as clear and precise as is necessary, and as is possible within the inexactitude inherent in the geological conditions prevailing on most sites. The Report is not of great value unless it provides exactly the information needed for the construction proposed.

In its most straightforward form, suitable for the usual study of foundation problems, the Report should consist of at least the following sections:

(1) *Detailed borehole records.* These should not be a copy of the driller's logs written on the site, but a carefully annotated record by an experienced soil engineer working from jar or bulk samples in the laboratory. The description of the strata passed through must be clear, accurate and unambiguous.

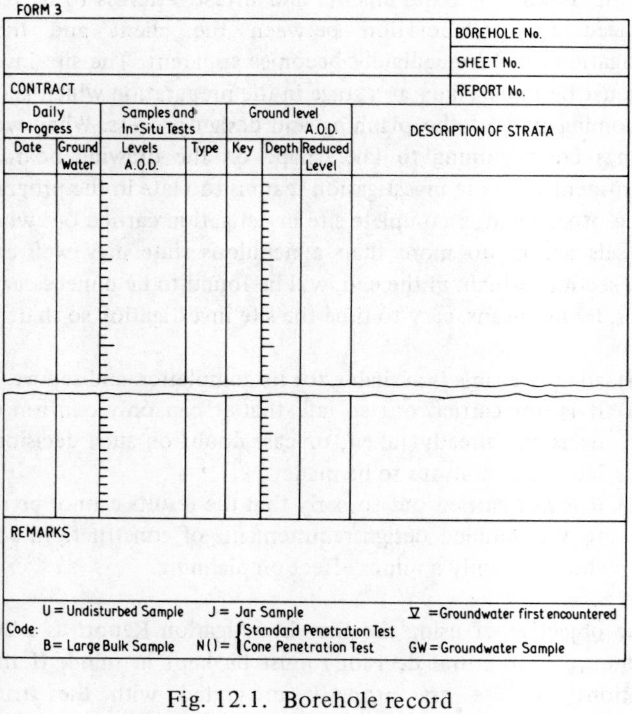

FORM 3

		BOREHOLE No.
		SHEET No.
CONTRACT		REPORT No.

Daily Progress		Samples and In-Situ Tests		Ground level A.O.D.			DESCRIPTION OF STRATA
Date	Ground Water	Levels A.O.D.	Type	Key	Depth	Reduced Level	

| REMARKS |

| Code: | U = Undisturbed Sample | J = Jar Sample | ▽ = Groundwater first encountered |
| | B = Large Bulk Sample | N( ) = { Standard Penetration Test / Cone Penetration Test | GW = Groundwater Sample |

Fig. 12.1. Borehole record

Water levels must be recorded, not only when first encountered, but also when finally settled down to equilibrium conditions. In clays, for example, it may take several days for a steady water table to be established. Boreholes must be left unfilled or fitted with standpipes if full value for the money spent on drilling is to be obtained. The location and movements of the water table are some of the more important data extracted from borehole records.

If the dip and strike of the strata over the site are to be recorded with any sense of reality, it is essential that the ground level at each borehole is recorded with reference to Ordnance datum or to a temporary site datum (AOD 'above Ordnance datum' or AD, 'above datum'). It is not sufficient to record depth below the surface to some specific feature, such as a change in strata. The depths below ground level to a particular stratum may be widely different in different boreholes, yet when levels are plotted to Ordnance or other datum, the stratum is likely to be found to have a well defined position and dip.

Later, when the results of *laboratory* and *in situ* tests are being considered, it should be possible to locate rapidly the position and Ordnance level to which any test applies. On the borehole record there must be full annotation of the levels at which samples have been obtained for testing or *in situ* tests have been made. The usual specification is that the properties of the soil should be examined either at every 1·5 m (approximately) or at change of strata.

(2) *Results of laboratory and in situ tests.* In the interests of efficient and timely reporting, only those tests should be made which are required for the particular conditions prevailing on the site. The indiscriminate application of all possible testing procedures to all available samples is deprecated. Unless the test gives information directed to the design in question, it is not required. Classification tests, especially in the secondary category for cohesive materials, are of value in predicting behaviour, but to present grading analyses of all granular soils encountered is not helpful. Consolidation tests are time-consuming; unless the proposed structure is likely to be sensitive to settlement and there is a substantial depth of cohesive soil, such tests will be of minor value. Similarly, the coefficient of compressibility is of more importance on normal sites than the coefficient of consolidation.

The aim of the investigator should be to match the tests to the conditions of the site and the requirements of the structure. Both of

301

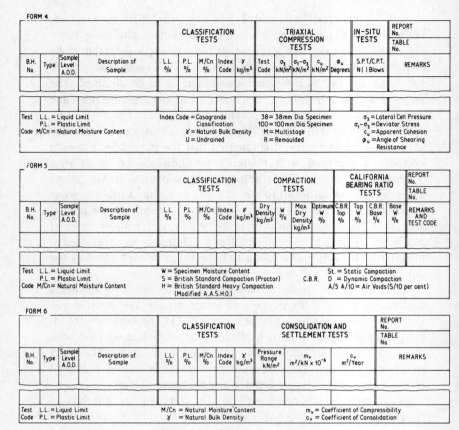

Fig. 12.2. Test results

these criteria must be kept in mind simultaneously when a testing schedule is being planned. Speed of reporting is valuable; time used in unnecessary testing detracts from the value of the report, even if it is a massive volume when completed. Excellence is not correlated with weight!

(3) *Profiles of strata.* In order to obtain the best interpretation from the results obtained, it is advisable to draw profiles of the underground strata by showing the sequence of strata in each borehole to Ordnance or other defined datum, perhaps with

annotation of test results against each stratum. When such diagrammatic representation of the underground conditions are related by spacing the boreholes in scaled sequence, a *profile* is produced. The horizontal scale is very much less than the vertical, so the gradient of a stratum along any chosen line of boreholes is much exaggerated. By linking boreholes in various directions across the site, in successive profiles, a good three-dimensional impression of the prevailing conditions can be obtained. This is a useful interpretative technique, perhaps not sufficiently used.

(4) *Interpretation and advice.* The concurrence of all the various aspects obtained must be interpreted by an engineer with considerable experience of site investigation. The young engineer to whom this volume is directed should take every opportunity of accompanying and listening to someone who has had a long experience of this type of work. It is certainly possible for even the inexperienced to apply the various methods of calculation described in the preceding chapters for such results as bearing capacity, settlement, stability and others, but such calculations can be misleading in the complex conditions often prevailing on a site, especially one of large dimensions. On a site with heterogeneous strata and materials, personal experience must be used as a catalyst to bring into true focus the various calculated results from tests and their applications. The conditions surrounding a site investigation are so uncertain of definition, that it is dangerous to rely on a few tests and calculations alone. Knowledge of what has occurred on other sites is of abiding value.

In interpretation, the engineer cannot be so definite in his predictions as to be unchallengeable. Conditions in investigations in soil mechanics are not sufficiently definable. He must, however, be as definite in his conclusions as is possible. The Report is not of great value if each of its conclusions is hedged about with qualifications to such an extent that it is difficult to determine what is being recommended. Every condition must be considered and related; a final conclusion as to stability, settlement or allowable bearing pressure should be made in clear terms. As an example of the need to consider all factors and not merely test results, there could be a condition where the water table is high. It is difficult in these circumstances to construct a large structure, perhaps with basements, without making some alteration to the pore-pressure conditions. Such alteration in the water table can have pronounced results in certain soils. The results of the originally reported tests may not refer to conditions after excavation has commenced. Such change must be anticipated and noted by the writer of the Report. For some reports, advice on construction may not be required. Such reports need be only factual, with borehole data and test results tabulated. The client, an engineer, is then qualified to make his own deductions. In spite of this, such a Report might well contain comments on observed conditions which would influence the proposed construction.

CHAPTER 13

In situ testing

Certain facts about the properties of the soil and all the information about the prevailing pore-pressure conditions can be obtained only by *in situ* testing. After samples have been removed to the laboratory for study, the engineer is confronted by the necessity of extracting further information from the mass of soil on which his structure is to rest. *In situ* testing cannot be carried out with the degree of accuracy or under the controlled conditions of laboratory testing. Some information cannot, however, be obtained in any other way than by inspecting the soil mass, and other facts can be obtained more rapidly by such inspection.

13.1 MEASUREMENT OF DENSITY

The *in situ* density of the soil is one of the most important properties required in soil mechanics investigations. Several methods of measuring the bulk density are detailed in B.S. 1377. From the bulk density and the moisture content the dry density is easily obtained.

Undisturbed sample method
For clay which is not too hard or stony, the simplest method is to employ a core cutter 100 mm in diameter and 130 mm deep. The tube is rammed into the soil, dug out, trimmed and weighed. The volume is determined from the dimensions of the cutter.

Sand-replacement method
This method is suitable for all soils but the apparatus is of a larger size

when coarse-grained soils are being tested. The apparatus used is a *sand-pouring cylinder.*

A cylindrical hole about 100 mm diameter and up to 150 mm deep is excavated in the soil to be tested. This is done through a circular hole in a metal tray which is used to collect the excavated material. This material is carefully weighed.

The sand-pouring cylinder is then set over the hole and sand is allowed to run from it into the hole. The loss of mass of sand in the sand-pouring cylinder (adjusted to allow for sand which ran through the shutter but did not enter the hole) allows of determination of the volume of the hole in the soil. Calibration is by means of a cylinder of known volume into which sand from the sand-pouring cylinder is allowed to flow.

Since the density of the sand used is sensitive to changes in moisture content, the sand is allowed to reach an equilibrium with the atmospheric humidity *before* the test is carried out. Oven-dried sand would not retain its original measured density if allowed to absorb moisture *during* the test.

A comparison of the weight of soil from the hole with the mass and therefore volume of the sand allowed to fill it, allows of the calculation of the bulk density of the soil. Details of the techniques used in this and other British Standard tests for soils are described in B. S. 1377.

Other methods. In very impervious soil the volume of the excavation can be measured by filling up with oil from a graduated vessel.

Water is sometimes used with a thin rubber bag, but this method is not very accurate because of the difficulty in making the bag fill the cavity completely.

In some laboratory tests it is necessary to measure the density of small samples, and for this purpose the volume is best determined by the displacement of mercury.

Bulk density from nuclear emissions

This technique, which has been studied in many countries, and for which commercial equipment has been developed, depends on the measurement of the energy of radioactive particles after they have passed through a mass of soil. The particles can be introduced into the soil either from a small source inserted at the end of a probe, or by emission from a source set in contact with the surface.

The first of these, known as the 'direct' method, uses gamma radiation to determine the bulk density of the soil. A caesium source at the end of a probe is inserted into the ground. During the test the intensity of the stream of gamma particles is measured as it emerges at the surface after passing through about 200 mm of soil from the buried source. The bulk density varies inversely as the intensity of gamma particles reaching the recording Geiger-Muller tube.

The second technique is used to determine the moisture content by means of a stream of fast neutrons. Radiation from a radium/ beryllium source at the surface is transmitted downwards into the mass of soil and is scattered back after having penetrated several inches. By 'backscatter' measurement of neutrons which have been slowed down by their passage through the soil molecules, the moisture content can be found.

These two tests are used separately to give a measure of the bulk density and of the moisture content. The dry density, which is of greater significance, can then be determined by the methods of Chapter 2.

These techniques are clearly of value in the control of compaction in embankments provided the accuracy of determination is sufficiently high. Such modern complex techniques are continually under review and development, and the meagre accuracy which radioactive methods, first showed increases with experience. Users of these methods, however, must be assured of the volume of soil to which the readings refer, and of what influence density and moisture gradients have on the results. One of the important features of these techniques is that, unless the original radiation is of a high intensity, the calibration of the apparatus is affected by the type of material whose density and moisture content are being measured.

Density from penetration tests
In sandy and gravelly soils and in weak rock, where the extraction and testing of 'undisturbed' samples is difficult and unreliable, the quality of the material can be judged and given a value from its resistance to the penetration of standard tools. There are two principle types of penetrometer: the *dynamic* where the penetration is effected by driving the tool into the stratum being tested, and the *static*, where the tool is pushed into the stratum without impact. Many methods of determining the density or packing of granular soils by penetration or 'sounding' tests have been devised since the early years of the

twentieth century. These vary from the simple hand-driving of a crowbar to the relatively elaborate equipment devised by the Dutch for the static testing of their soft sediments. Of these methods, three are found in general use today. They are:

(i) *The Standard Penetration test* in which a tube (originally intended as a sampler and sometimes called a *spoon*) is driven by blows of a falling mass, the energy of which is standardized. The tool is open at the end and can be split lengthwise; it might thus contain a sample when retrieved, but its principal use is to determine the density of the soil.

When a borehole has been cleaned out to the depth at which a test is to be made, the *penetrometer* is lowered, and driven 150 mm into the soil by standard blows. It is then driven a further 300 mm and the number of blows required to do this is internationally known as N, the SPT number, or the *Standard Penetration Test* value. The blows are delivered by a 65 kg mass dropping through 760 mm. (The peculiar values of these parameters is due to the original test being carried out in Imperial units; 140 lb through 30 in.)

(ii) *The Dynamic Cone Penetration test.* When driving through gravel rather than sand, the open end of the SPT spoon is replaced by a solid cone. The results are given as the CPT number. The cone has a 60° angle and it is customary to assume that the SPT and CPT values are the same for gravels. In sands, the CPT results are somewhat higher than for the SPT, but in most surveys the SPT would be used in sand and not the CPT which is usually restricted to denser sediments.

(iii) *The Static Cone Penetrometer* is often referred to as the 'Dutch Cone Test'. This time, the cone is pushed into the soil and not driven by blows. It is not suitable for hard strata or gravels, but was developed by the Dutch for their conservancy engineering in alluvial and deltaic conditions. It is thus at its best in sands and silts. The penetration rate is from 15 to 20 mm s^{-1}, and motor or engine power is required as well as substantial anchorages to take up the vertical reactions. These reactions can be supplied by screwed anchors.

By the study of test results in known conditions, engineers have gradually developed some correlation between N values and more specific properties of the soil. Some attempt has been made to link bearing capacity and settlement to the N values, and these are discussed elsewhere in the book. The chief relationship between N and another property of the soil is with density. The relationships obtained can be considered reasonably reliable so long as non-cohesive soils are encountered. As the cohesive end of the soil classification is approached, the relationships become relatively less reliable, and are more influenced by pore-water conditions. For silts Menzenbach found that 0·85 was the upper limit of the degree of saturation for which correlation was possible. With saturation above this figure, the scatter became too great to be acceptable.

For non-cohesive soil an estimate of the relative density can be made from Table 13.1, but these figures are only a guide. The depth of overburden above the level at which the test is taken may obscure the significance of the N values.

13.2 MEASUREMENT OF BEARING CAPACITY

The *in situ* measurement of the strength of soil can clearly apply only to the surface or near-surface strata, or to measurements taken in boreholes. It might seem at first glance that tests taken on the ground *in situ* would give a closer estimate of strength than methods requiring the extraction of so-called 'undisturbed' samples, tests on these and calculation. *In situ* tests, however, have their own disadvantages, and the results obtained are not indubitably better or more accurate than those obtained from samples in the laboratory.

Field loading tests

Direct loading tests in the field, intended to determine the ultimate bearing capacity of the soil (and, incidentally, of probable settlement), are not only time consuming; they are also expensive. They occupy valuable space on the site, for such tests must be carried out where the foundations are ultimately to rest. They should be used only when it is quite clearly impossible to obtain a reliable estimate of bearing capacity by laboratory methods or by *in situ* tests of a less elaborate nature. Sometimes, loading tests are the only possible method; they are particularly relevant in difficult made-ground. A paint factory wished to extend its workshop, and the ground consisted of brightly-coloured and dense layers of long-dried-out paint wastes to a great depth! A field loading test successfully gave the value of the loading which should not be exceeded. The field loading test is the last resort. It is applicable when the zone of the soil within which stresses will be critical (when the structure is completed) lies close to the surface.

Several plates of different dimensions are loaded at representative sites, and their settlement measured carefully. If the test is made at the bottom of a trial pit, it is important that the sides of the pit should be well removed from the plate itself. This has the effect of eliminating the influence of the overburden represented by the depth of the trial pit

The chief sources of error in estimates of settlement made from loading tests are:

(i) the difference in area and possibly in shape between the loaded plates and the full-sized footings;

(ii) the greater depth of soil affected by the structure as compared with the test plate. Loading tests give little information regarding differential settlement, which is so often of vital importance.

Load/settlement curves are obtained from the behaviour of the plates under increments of load. The ultimate bearing capacity for each is represented by that portion of the curve where the settlement increased rapidly with only a slight increase in loading. The closer the size of the largest bearing plate to the size of the proposed footing, the more accurate will be the bearing capacity estimated. Extrapolation from the results of three or more tests can give some indication of the ultimate bearing capacity of the foundation, but it should be qualified and guided by experience.

Plates used should be as large as possible, but the disadvantge of the *Field Loading Test* is shown by the fact that it is usually uneconomic to use plates of a satisfactorily large size. A plate approaching a metre square needs a very large load to produce failure, but to try to make a true estimate of bearing capacity for the final structure from small plates is to delude oneself. For the test to be successful, it is important that the final excavation to test level should not be done until all arrangements for loading are completed and ready to be applied.

Borehole loading tests
Loading tests at the surface or at the bottom of a trial pit influence the soil only to a shallow depth, perhaps only to twice the width of the plate used. As many large structures are founded at much deeper depths, this gives insufficient information. Accordingly, the practice has grown, for large and important works, of testing the foundation

soil and rock at various horizons within the boreholes driven during the site investigation. This has the advantage of testing quite undisturbed soil at the levels to be used for the proposed foundations, and of automatically including an allowance for the weight of the overburden. Measurements of penetration against load are made by equipment built into the loading column. The disadvantages of this type of testing are (i) that the diameter of the loading plates are restricted to less than the diameter of the boreholes, and thus cannot affect large volumes of soil, and (ii) that the bedding of the plate at the bottom of the hole cannot normally be inspected. At the time of publication, the largest diameter of borehole used for such loading tests has been 900 mm, with loading plates of 865 mm diameter. These have been used by the Building Research Establishment (A. Marsland). Such diameters allow of men being lowered to prepare the bedding of the plate, and the larger diameter gives a better estimation of the properties of the soil at depth than could be obtained in any other way. Borehole loading tests require similar, but heavier anchors (maximum load so far, 500 tonne) to those used in the surface loading tests.

California bearing ratio

This is a type of plate bearing test with a very small 'plate'. It is usually carried out on samples in the laboratory, and is described in the next chapter. It can, however, be made *in situ,* the reaction to the pressure applied being taken by the weight of a vehicle. The test gives an empirical result which is then applied, equally empirically to the design of the thickness of the construction of flexible road pavements.

Vane test

With soft clay where it is difficult or impossible to obtain a sample for laboratory test, the shear strength can be obtained by *in situ* testing and can be used to find the bearing capacity. This is a type of penetration test, for a tube carrying the testing tool is driven by usual boring methods to the required depth, and the testing tool then made to penetrate into the layer to be investigated. The tool consists of a vane comprising four thin rectangular blades. The vane is generally twice as deep as it is wide. Small shear vanes can be obtained individually calibrated and reading directly in values of shearing strength rather than of torque. Larger vane borers may be designed to be driven into the ground.

Vanes, as shown in Fig. 13.1 generally have blades 150 mm deep by 75 mm wide (across the diameter) for soils of shear strengths up to 50 kN m^{-2}. For stiffer clays, up to 75 kN m^{-2}, the corresponding dimensions are 100 and 50 mm.

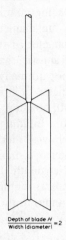

Fig. 13.1. Vane test

Two methods are used to drive the vane to its required position. The vane may be lowered to the bottom of a cased borehole with no protection, and then driven into the soil by at least three diameters of the borehole. Alternatively the vane may be driven from the surface to the selected stratum. It is then required to be protected. Protection takes the form of an enclosing jacket or cartridge with a driving shoe. The vane is released from the protection and driven for at least 0·5 m before the test is made.

By means of gearing at the surface a torque is applied to the boring rod carrying the vane, which is thus twisted in the soil. A cylindrical surface of rupture develops, and, assuming that shearing resistance is distributed uniformly over this surface, the shear strengths of the clay is given by the expression:

$$M = \pi S \left(\frac{D^2 H}{2} + \frac{D^3}{6} \right).$$

M is the torque applied, D is the diameter or breadth of the blades and

H is the height of the blades. Or, when

$$H = 2D$$

this becomes

$$S = \frac{6M}{7\pi D^3} = \frac{M}{3 \cdot 66 D^3}.$$

The terms in this expression are assumed to be evaluated in N and m units, giving S in kN m^{-2}. It is more convenient, however, to measure M in Nm and D in mm.

The expression then becomes

$$S = \frac{M \times 10^6}{3 \cdot 66 D^3}$$

when M is in Nm, D in m and S in kN m^{-2}.

The torsional strain is applied at a uniform rate, usually $0 \cdot 1$ to $0 \cdot 2°$ s^{-1}, and the ultimate torque is measured by a suitable torque meter. A smaller form of this apparatus is used in the laboratory.

Allowable bearing pressure from penetration tests

The Standard Penetration Test (SPT) or the Cone Penetration Test (CPT) or other tests giving the relative resistance of the strata to penetration can be used to predict not only the density of packing of the material, but also the bearing pressure. This is usually determined as the *allowable bearing pressure* (ABP) and not as the ultimate (as obtained from bearing capacity factors). The word 'allowable' requires definition. In this study of granular material (to which the SPT particularly applies) the criteria applied to the allowable bearing pressure are:

   (i) under this pressure the soil does not show any sign of failure;
   (ii) the settlement produced is not more than 25 mm;
   (iii) the soil considered is a dry or merely moist sand.

Table 13.1 shows three properties which have been correlated approximately with the SPT value. N is the number of blows per 300 mm of penetration when the test is carried out in the standard manner.

The bearing pressure can hardly be stated with certainty or

unassailable accuracy. The figures given in Table 13.1 are guides, and perhaps only approximate guides. Rough approximations can also be made by estimating that the allowable bearing pressure lies between $11 N$ and $13 N$ kN m^{-2}. This is probably much too approximate, but can be easily remembered.

Table 13.1 *Guiding and approximate correlations between penetration resistance and other soil properties*

SPT value (blows per foot) (blows per 300 mm)	State of packing (or density)	Angle of shearing resistance (°)	Approximate allowable bearing capacity (kN m^{-2})
< 4	Very loose	< 29°	< 50
4-10	Loose	29-30°	50-120
10-30	Compact	30-35°	30-370
30-50	Dense	35-40°	380-550
> 50	Very dense	> 40°	> 550

When footings must be designed, several physical features on the site must also be brought into consideration. This aspect of the use of Table 13.1 is fully discussed in a previous chapter.

13.3 LOADING TESTS ON PILES

The purpose of a loading test may be either to determine the ultimate bearing capacity of a pile or to check whether a pile can safely carry its design load as calculated by one of the pile formulae. In the latter case the loading should be carried up to at least 50 per cent above the design working load to ensure a reasonable factor of safety.

Test loads may be applied in the form of kentledge placed directly on a platform on top of the pile, or indirectly at the end of a lever. Alternatively, the test pile may be driven in by a hydraulic jack bearing against a rigid beam supported on two firmly driven anchor piles. The anchor piles should be at least 2 m away from the test pile.

Maintained load method
The usual method of procedure is that described in C.P. 2004. Similar procedures are specified in American and continental codes. The test load is applied by equal increments, the load at each stage being

maintained until all observable settlement has ceased. Records are kept of time, load, and settlement. With friction piles there is usually a clearly defined peak load indicating the ultimate bearing capacity (*A* on curve 1, Fig. 13.2). When the principal resistance is the end-bearing capacity, the load/settlement often becomes tangential to a straight line representing a small uniform increase of settlement with increase of load. The load at the point of tangency should be taken as the ultimate load (*B* on curve 2, Fig. 13.2). A factor of safety of $1\frac{1}{2}$ to 2 should be applied to determine the safe working load.

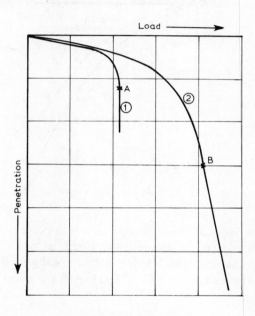

Fig. 13.2. Test on piles

When a load is applied to a pile the resulting settlement is made up partly of elastic compression of the pile and partly of the plastic deformation of the soil. It is sometimes useful to determine the ratio between these two deformations. In Fig. 13.3, *OL* is a load/settlement curve obtained from a maintained loading test. If each increment of load is allowed to act until movement has ceased and then removed before the next increment is applied the permanent set or plastic settlement can be measured. Two such releases of load are shown in

Fig. 13.3, where the distances from the origin to X and Y represent the permanent settlements for the two loads concerned. If an ordinate is raised from all points such as X or Y equal in height to the load on the pile before release, a curve of permanent settlement/load can be drawn. At any load, such as is indicated by P, the elastic compression of the pile is shown as ML, the intercept between the curves of total and plastic settlement.

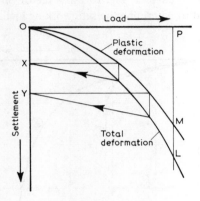

Fig. 13.3. Load–settlement curve for pile

From this value of elastic compression it is possible to make an approximate estimate of the effective length of pile l which is in action under the load. This length is not necessarily the whole length of the pile. The relationship defined by Hooke's law gives

$$\frac{R/A}{\delta/l} = E$$

$$l = \frac{E/AS}{R}$$

The resistance of the soil to the applied loading may be assumed to originate from the depth defined by l. When this depth is compared with the soil profile obtained during the site investigation the strata which are concerned in developing resistance can be identified.

Constant-rate of-penetration method
The pile is forced into the ground at a uniform rate of settlement and

the resistance is continuously measured. The speed should be chosen so that the duration of the test is similar to that of an undrained shear test in the laboratory. The exact rate of loading is not critical, but it is essential that the loading should be uniform and free from jerks. A rate of penetration of 8 mm min^{-1} is suitable for friction piles in clay, but this should be increased to 15 mm min^{-1} or more for end-bearing piles in sand or gravel. The general shape of the load/penetration curve is similar to that obtained from the maintained load test, and the ultimate bearing capacity is found in the same way.

By this method a great deal of time is saved compared with a maintained load test, and it is a comparatively simple matter to repeat the test, if necessary, at several depths of driving in order to arrive at a an economic design.

Interpretation of loading tests
The data obtained by careful study of the load/settlement diagrams furnish a reasonably accurate estimate of ultimate bearing capacity in cohesionless soils. In cohesive soils, however, the data must be treated with caution. These load/settlement diagrams relate only to short-term loading, hence the settlement which occurs under a pile carrying a static load for a long period cannot be estimated accurately from a loading test. From the immediate intensity of loading applied to the soil it is possible to estimate probable final settlements by the methods of Chapter 9, but the deformation measured during the loading test are by themselves no sure guide.

When the ultimate bearing capacity is not clearly shown by the load/settlement curve obtained from either a maintained load test or a constant-rate-of-penetration test, rule of thumb methods are often used to estimate the safe load on the pile. Such approximate and empirical rules do not fall within the scope of soil mechanics. One common empirical rule may be mentioned: the assumption that the ultimate bearing capacity is fully mobilized when the settlement is 10 per cent of the diameter of the pile. This rule obviously does not apply to large bored piles.

13.4 MEASUREMENT OF PORE-PRESSURE

The significance of pore-water pressure (or pore-pressure if we assume that the pressure of air in the soil does not confuse the value) is discussed in other chapters. More effort has been made in recent years to record pore-water pressure accurately, not only during the site investigation, but perhaps for months or years after construction is complete. For example, pore-pressure in dams must be monitored during and after construction to ensure that there is no dangerous reduction in the effective stresses, and that the pore-pressures are gradually dissipated. Under foundations the presence of water must be established to determine the precautions to be taken in the design of basements or buoyant foundations to resist the effects of water pressure.

In determining the presence or absence of water pressure, no reliance can be placed on the appearance or non-appearance of water in a borehole during or immediately after drilling. This applies particularly in clays, where a hole may be bored quite dry, but show a build-up of water pressure over succeeding days. The fluctuation of pore-pressure (perhaps near a tidal river) is also of importance. Borehole records are satisfactory only if they show variations of water level with time. Unfortunately, site investigations are often commissioned too close to the date of construction to allow of an accurate assessment of variation in pore-pressure, although this is sometimes of greater import than bearing pressure and other properties much more carefully analysed.

Standpipes

On any site investigation one or more boreholes should be provided with a *standpipe,* which is a tube, perhaps of 50 mm diameter, lowered down the borehole and packed round with sand or clay. It has a perforated portion at the lower end to which water flows and rises in the tube to the *water table* level. An open borehole without a standpipe is a cruder method of obtaining a similar result. The level of the water in the standpipe is read by lowering a metallic plumb-bob on the end of a coaxial cable. When the plumb-bob touches the water surface an electrical circuit is closed and an acoustic signal heard. It should be remembered that there is always a time-lag, sometimes quite long, between the setting up of a measuring system and the recording of the true pore-water pressures.

Piezometer installations

Strictly speaking, an open borehole or a standpipe is a *piezometer* which merely means a *pressure measuring device,* but the word is usually restricted to describe the apparatus used in more complex and more accurate methods of pressure measurement. The standpipe gives a general assessment of the water level and so of the pore-pressure at any depth in the borehole, and that may be all that is required. However, in an open borehole, and even with a standpipe, the pressure recorded results from the filling of the borehole with water from all the strata through which it has passed. Conditions such as perched water tables or variations in pressure which are not dependent on general conditions do not appear in the results obtained.

If the pressures in the pore-water at particular points in a soil mass are required, measuring devices at these points are isolated from the disturbing effects of other nearby water pressures. In a borehole this is done by putting an impervious plug above and below the measuring instrument, and surrounding it with a pervious medium such as sand. The measuring instrument is then connected to the surface in such a way as to provide a reliable signale which can be translated into terms of pore-water pressure.

The chief devices used are porous tips which allow the water to flow up a tube to a rest level, pneumatic recorders which balance the pore-water pressure against air or oil pressure, and electrical recording from diaphragms actuated by the water pressure. Of these, the porous-tip piezometers are the most commonly used. They have been developed into effective systems of measurement and recording.

Piezometer tips and tubes

The *piezometer tips* which are porous and in contact with the water under pressure have been the subject of much experiment. They may be solid or hollow, the latter predominating, and may be made of ceramic stone or of the less corrodible metals (bronze or brass). The pores in these tips are very small and must be protected from being sealed by the invasion of clay or silt particles carried by the water. This is done by providing a permeable volume of sand around the tip of such grading that small particles are prevented from migrating.

We thus have a porous cylinder, or sometimes (in earth masses) a disc, to which only the localized water pressure has access. This pressure is then transferred (to a position where its value can be recorded) by allowing the water to flow in a tube towards its rest level. It is essential that the tube should be able to withstand abuse since it is laid within a borehole or in a trench and subjected often to physical disturbance. It must also be impervious, since only the water from the tip should exert pressure through the tube. Experiments over the years have produced a nylon tube coated with polythene as the best answer to such a specification, and these are now almost universally used. In most piezometers, twin tubes are used to connect the tips to the surface so that water may be circulated through the piezometer tip, as a precautionary and preparatory measure.

Elimination of air

The object of a piezometer installation is to measure the pore-water pressure. If the soil (unsaturated) or the water contains air, the pressure of the air obscures the true value of the water pressure. It is important to remove the air or, if that is not possible, to eliminate its effects. To eliminate air bubbles in the water within the tip and the impervious tubing is relatively simple. A volume of water is placed under vacuum and the air in it removed. This de-aired water is then forced down one of the nylon tubes to the piezometer tip and up the other one to the de-airing unit at the surface. Circulation of water through the tip is continued until the whole of the volume of water in tubes and tip is air-free. The method of forcing the de-aired water into the tubes and circulating it must be done without access to further air. It is usually accomplished by expanding a rubber bag by pumping tap water into it. Since this bag is submerged in a closed vessel containing the de-aired water, its expansion forces the water into the piezometer system and back through the second tube mentioned above.

Although it is possible to eliminate air from the water used to transmit pressure to the recording positions, air in the soil cannot be eliminated. It is important, therefore, that none of the air in the soil can penetrate through the pores of the piezometer tip and so invalidate the values of the pore-water pressures measured. This is done by decreasing the pore sizes in the walls of the piezometer tip to such an extent that the excess of air pressure over water pressure must be high before air can enter the system from the soil. The tip is then said to have a *high air entry value*. The pressures then recorded may be assumed to be those of the pore-water pressure only, especially if the de-airing process is repeated at intervals to ensure the continued absence of air from the water in the system.

Recording the pore-water pressure

When the water in the piezometer system has been de-aired, one of the twin tubes is closed and the other used as a pressure tube to carry the pressure of the water at the tip to the rest level it is attempting to reach. This rest level may be below the surface, when its position can be recorded with a *dipmeter* as in a standpipe. Piezometer installations are usually set up where there is reason to believe that the rest level of the water will be above the surface, or can be made to appear above the surface by suitable arrangement of the location of the recording apparatus.

Various methods of recording pressure as used in industry can be adopted (e.g. the Bourdon gauge), but there is seldom any advantage to be gained in inserting what is a possible source of error into the system. The de-aired water transmitting the pressure is usually fed into a mercury manometer system such as is used in measuring pressure in laboratories. The pressure in metres of water head is then the difference between the level of the piezometer and that of the manometer, plus the head recorded by the manometer translated from head of mercury to head of water. A back-pressure or header tank is usually provided so that the displacement of the mercury manometer can be controlled. When a header tank is provided, the pressure is then the difference between the level of the tip and that of the header tank plus the water head recorded by the manometer. Installations requiring the measurement of pressure from several piezometers are not uncommon on important works, and are installed in special buildings which may be used for a long period after construction of the engineering work is complete.

Although hydraulic recording of pressure by manometers is a commonly adopted method, it does not respond quickly to changes in pressure. When such rapid changes are expected, it is worth considering some of the advances which have been made in the direction of more flexible recording and a faster response to changes in pressure. A single pressure transducer can be used to scan the readings of up to 20 piezometers in sequence, the recording being made on a chart. The use of diaphragms sustaining the pore-water pressure on one side and air pressure or some system of electrical recording (such as the vibrating wire) on the other eliminates some of the problems of hydraulic recording; a more rapid response time can be registered.

13.5 MEASUREMENT OF PERMEABILITY

Laboratory determinations of permeability are often unreliable because of the difficulty of ensuring that the degree of compaction of the soil in the permeameter is comparable with that in the ground. For large engineering projects it is often advisable to measure the *in situ* permeability of the soil by means of pumping tests.

Pumping tests
Fig. 13.4 shows sections through a well, the curved line indicating the profile of the water governed by the permeability of the soil and the slope of the original water table.

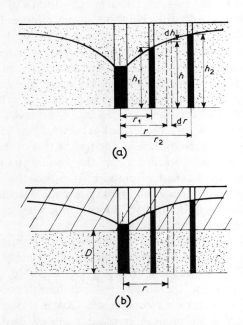

(a)

(b)

Fig. 13.4. Pumping tests for permeability

In Fig. 13.4a the rate of radial flow through an element at radius r is given by

$$q = Aki = 2\pi rhk \frac{\mathrm{d}h}{\mathrm{d}r}$$

$$\frac{\mathrm{d}r}{r} = \frac{2\pi kh\mathrm{d}h}{q}$$

Integrating between r_1 and r_2 and between the corresponding heights h_1 and h_2

$$\log_\epsilon \frac{r_2}{r_1} = \frac{\pi k}{q} (h_2^2 - h_1^2)$$

Therefore

$$k = \frac{q \log_\epsilon \dfrac{r_2}{r_1}}{\pi (h_2^2 - h_1^2)}$$

Fig. 13.4b shows an artesian well in which the height of the permeable stratum D is constant. The integration of the differential equation gives

$$\log_\epsilon \frac{r_2}{r_1} = \frac{2\pi k D}{q} (h_2 - h_1)$$

$$k = \frac{q \log_\epsilon \dfrac{r_2}{r_1}}{2\pi D (h_2 - h_1)}$$

The data for the determination of k from the above equations are obtained as follows. A test well is drilled to the bottom of the permeable stratum, and water is pumped out at a steady rate until the level in the well becomes practically steady. Several observation boreholes or wells are made around the test well, in which the level of the water table is observed. If the general direction of ground-water flow is known, the observation boreholes should be arranged in two straight lines, one along and the other at right angles to that direction.

In situ permeability by piezometer

Since the modern piezometer is fitted with twin tubes, any maintained difference in pressure in these tubes causes an outflow of water from the piezometer into the surrounding ground. The rate of this outflow is dependent on the coefficient of permeability of the soil. Thus, knowing the annular area through which the flow takes place from the piezometer or its surrounding cylindrical filter, together with the constant head causing flow and the volume of water lost into the soil, the coefficient of permeability can be calculated.

To obtain data for this calculation requires a manometer board adapted to give values of the constant pressure maintained on the piezometer tip, and of the resulting flow of de-aired water into the soil. The first of these is measured by the usual mercury manometers and maintained either by a constant-head supply tank or by a mercury pot constant-head unit. The second value is measured as the volume of water lost in a measuring-burette system, where the movement of either an air/water or a paraffin/water interface gives the rate of flow into the soil.

The measurement of pressures by piezometer does not require movement of water during recording but, in measuring permeability by piezometers, there is a danger of loss of head due to hydraulic resistance to flow, especially if the tubes from the piezometer tip to the manometer board are long. If the piezometer is not located in a sand filter but pushed directly into the soil (e.g. in soft estuarine deposits), there is also the danger that the fine particles of the soil may smear the surface of the piezometer tip, restricting flow and altering the permeability from its true value. Furthermore, the problem of the presence of air, referred to in the study of pore-pressure above, also occurs here. It is, however more serious in its effect on accuracy. In the recording of pore-pressure it was necessary only to keep the air in the soil from entering the measuring system through the pores of the piezometer tip. In the permeability test the flow is outwards, and air trapped in the filter of sand surrounding the tip may falsify the value of the rate of flow. The sand for the filter must, therefore, be de-aired as was the water in the pore-pressure tests.

These are formidable difficulties, but if they can be overcome or allowed for, as various writers have suggested, then theories can be applied, linking the adjusted site readings to the value of the coefficient of permeability. These theories are complex, and not normally studied at the stage to which the reader of this book is supposed to have attained. It is worth remembering, however, that values of permeability (which lead to the estimation of consolidation under load) may be determined *in situ* if precautions are taken to eliminate sources of error in site operations.

13.6 GEOPHYSICAL METHODS

Geophysical methods comprise techniques of studying underlying rocks by gravimetric, magnetic, electrical and seismic procedures. The object is to determine, with as much detail as is possible, the configuration of at least the firmer strata underlying the site. Errors have less effect if a greater depth is explored. For engineering sites the depths are less, and the methods employed, if they are to be effective, should provide a high resolution, and the possibility of converting readings to quantitative rather than merely qualitative estimation of conditions. This is a difficult specification; the accuracy of such observations are militated against by the greater likelihood of error in shallow exploration. For these reasons, the methods employed for foundation problems have generally been those of *electrical resistivity* and *seismic reflection* and *refraction*. For the location of suspected sub-surface cavities or old mine shafts, gravimetric and magnetic methods have also been used.

The checking of ground conditions between boreholes by these methods assists in obtaining a clearer understanding of the lie of the strata. Areas of special significance are revealed so that check boreholes can be located to the best advantage. The best position for a particular structure can be selected rapidly, and a closer investigation made by traditional methods. This kind of decision must often be made for large dams where geophysical methods are of particular value.

Resistivity method

The electrical resistivity method is the most useful in site investigation for buildings or civil engineering structures. Differences in conductivity (inverse of resistivity) are large for different types of rock; variations of sub-surface strata can, therefore be reasonably closely identified by this geophysical method. Resistivity of a rock mass depends largely on the moisture content of the material. It is not a defined property of the mineral content. Thus, the changes in underground geological conditions are reflected only by comparative changes in resistivity. The figures obtained are not definitive for a particular kind of soil or rock.

The technique adopted in the resistivity method is to place four electrodes in line, at equal distances a from each other (three equal spaces). These electrodes are placed a short distance into the ground, and a current is passed between the two outer electrodes. The voltage across the two inner electrodes is then measured. A numerical value is obtained for each stance of electrodes by multiplying the spacing between two adjacent electrodes by the voltage across the inner ones. This value is then divided by the current (flowing between the outer electrodes) and multiplied by the outer electrodes) and multiplied by the constant 2π, to give the value of resistivity.

$$\text{Resistivity} = \frac{2\pi a V}{I}.$$

If the spacing of the electrodes is kept constant and the whole stance is moved across the site, a survey of resisitivity at constant depth is obtained. If the equal spacings between the electrodes are increased (while they still remain equal) and the centre of the system remains fixed, the survey of resisitivity probes more deeply at the same position. The depth reached by any reading is approximately equal to the spacing between adjacent electrodes.

With electrode spacing as abscissae and resistivity as ordinates, a curve can be drawn which gives a qualitative and partially quantitative picture of sub-surface conditions. If the resistivity increases with wider spacings, the indications are that materials of increasing resistance, such as rock, are found at deeper depths. If resistivity decreases with increasing spacing, the lower strata are likely to be of softer materials such as clays. Any sudden change in configuration of the curve shows a discontinuity in the strata, whose depth from the surface can be approximately determined. This perhaps indicates the presence of bedrock or of a geological fault. The resistivity method also offers the opportunity for locating gravel and sand deposits and for distinguishing between large boulders in boulder clay and the presence of true bedrock. This is a distinction which normal boreholes sometimes have difficulty in making without excessive chiselling.

Seismic method

When an explosion is caused within soil near the surface, waves of various patterns pass through the soil and are reflected and refracted from the interfaces of the different underlying strata. By a study of these waves and their times of travel from the shot to the detector, it is possible to map the sub-surface strata in sufficient detail to allow at least a qualitative estimation of the conditions to be encountered in foundation design. More detailed inspection by traditional methods is usually advisable, but the seismic method gives a clearer indication of changes in strata than does the more generally used lesistivity method.

Again, a graph is plotted, this time of the relationship between time of travel of the rays through the soil against distance between the point of the explosion and the wave detector. A discontinuity in the graph shows a discontinuity in the strata, and a change in slope shows a change in the strata.

In general, interpretation of geophysical surveys is not easy. They should be used only when it is impossible to obtain information by the normal method of borehole logging. The geophysical survey, however, gives a rapid means of carrying out a preliminary survey, after which the most effective positions for boreholes are more easily chosen.

To summarize, the seismic method will locate rock or similar hard strata more cheaply than by boring, and with sufficient accuracy to pinpoint areas requiring further investigation; a reduction in cost of the main survey is also possible. The resistivity method is the more useful on foundation sites. It can cover a large area quickly and help to fill in the area between boreholes. It is the method most easily carried out with the minimum of specific training.

13.7 TESTS FOR CONFIRMATION AND RESEARCH

In the usual range of construction it is not common to find follow-up tests being carried out for confirmation of the design or for the sake of research. If the building has obviously failed to carry out its functions, or if movement and cracks have appeared, tests and measurements will, of course, be set in hand. Research organizations such as the Building Research Establishment have been very active in research-testing, and have made many observations on construction sites and completed structures. Such bodies have the resources and staff to prosecute studies for the sake of gaining a clearer understanding of the problems of a particular type of construction, such as a dam or a tunnel or a buoyant foundation. Larger consulting firms have also monitored their constructions with the object of building up a body of knowledge for later use. Relatively few structures are studied in this way; results from many more would increase accumulated knowledge and produce further advances. It is not surprising that the smaller firms do not examine the performance of every structure they erect. The cost of such an enterprise is not limited to the purchase of instrumentation, but also draws on the time of staff to make and record the observations, perhaps for a considerable time into the future.

Before the engineer sets out on a programme of confirmatory and research tests, he must be quite sure what readings are essential, and the purpose for which they are required must be explicitly defined. He must be sure whether he needs readings taken during construction, as well as after the structure is complete. He must know if there will be a need for a continuing programme of monitoring the behaviour of the structure, and how long this should be extended. He must assure himself that access is provided to levelling and stressing points and that they are adequately protected, especially during construction.

Some of the types of soil tests to be carried out in this phase (for confirmation of predictions and the advance of knowledge) are listed at the end of Chapter 12. Some have already been discussed in the sections on design and control above. Those which still need consideration principally concern measurement of movement in structures or in a soil mass, and the measurement of earth pressures in dams, in foundations and against structures such as retaining walls.

Measurement of settlement

The measurement of the settlement of a structure and the speed at which this settlement takes place is one of the most commonly pursued programmes of testing. If the predictive methods of Chapter 9 are to be checked for their performance in practice, the settlement of the structure and of the ground nearby must be measured. This requires the provision of *settlement points* both on the structure and in the ground, and *datum points* in stable positions from which the relative movement of the settlement points can be measured.

The *settlement points* must be well fixed into the building and should have a cap screwed into the base (let into the stone or concrete of the building). This cap should be flush with the surface so that the apparatus cannot be damaged. When levels have to be taken, the cap is removed and a short machined pillar dropped into a machined hole in the base. The top of this pillar should be rounded so that a levelling staff can be used on it. Brass or other non-corrodible metal should be used, although hard steel settlement points are not unknown. Brass pins with rounded ends can also be set to project horizontally from a building, for the support of a precise levelling staff. If the building is in a remote area with little traffic or likelihood of being disturbed, the pin can be mounted permanently. Otherwise it should be screwed into a socket for each observation, the socket being protected by a cap between visits of the surveyor.

When it is necessary to study the settlement of the ground, the settlement point should be installed at some distance below the surface. This can be done in a borehole. The head of the probe is concreted in or bedded in sand near the bottom of the hole, but the major portion of the rod joining the fixed end to the upper levels should lie in a protective tube packed round with sand or other backfill. The top of the rod should be rounded and set some distance below the surface so that a cover can protect it from disturbance. Such covers should always be flush with the ground and incapable of being opened by the unauthorized.

The datum point which is, essentially, a benchmark, must be set up at some distance from the site of the measurements of settlement, and fixed in such a way as to eliminate any possibility of movement taking place. All datum point installations for the monitoring of settlement have several features in common. The reference point is fixed at some depth (up to 5 or 6 m, but normally somewhat less). The reference point consists of a plug set in concrete at the bottom of a borehole, and of such a type that movement within the concrete is not possible. This plug is then connected to the upper levels by a permanent (although sometimes detachable) rod which terminates in a rounded head some slight distance below the surface, but accessible to a levelling staff. The protective tube, as was noted for the settlement point above, is also provided, sometimes in telescopic lengths, in a further attempt to make sure that any movement near the surface, even of the protective tube, cannot be transmitted to the rod or the datum plug at the bottom of the borehole. Again, the whole is protected by a heavy socket and cap at ground level.

Modern developments have allowed of more complex measurement of movements within the ground and near the structure in question, before and after it has loaded its foundations. The use of the Building Research Establishment's *borehole extensometer,* for example, records the deformation of ground in these situations. This system records relative displacements, not only of one point on or near the surface, but at intervals throughout the depth of the borehole, by means of circular magnets and reed switches.

When it is not possible to bore vertically at the point where movement in the soil is required to be measured, a system using water levels can be used. A cell containing the end of a tube which can overflow to waste is embedded in a soil mass, and the other end of the tube is connected to a standpipe at some convenient point. Settlement of the cell within the soil causes a drop in level of the water in the standpipe consequent on overflow from the end of the tube in the cell. This is the simple principle on which various hydraulic systems are constructed for remote reading of settlement; the physical construction of the systems brings in other problems which do not affect the principle.

Measurement of lateral movement

Vertical settlement is not difficult to measure, but lateral movement presents problems in design and installation of apparatus and in interpretation. If the point whose lateral movement under load is required is easily visible, such as the top of a pile in a sheet pile wall, precise measurement by normal surveying techniques can give the figures required. If movements of points within a mass of soil, such as a dam, are the subject of investigation, the measurement of the lateral component is not so easy. The insertion of an object which will move with the soil mass would clearly be a solution if the location of the object when buried in the soil could be established.

At Scammonden Dam in Yorkshire the locations of buried plates were successfully established. The plates were set vertically and in line and were 300 mm square. Through a hole in the centre of each plate a long p.v.c. tube was inserted passing through several plates, and construction of the dam continued. The location of a plate was established at each reading of the lateral movement by passing an induction coil through the pipe by an ingenious pneumatic method. As the coil reached the location of one of the plates (which might have moved laterally since the previous reading) its inductance changed and the position of each plate was determined by knowing the distance of the coil from the entrance to the pipe. This is an elaborate method suitable only for large-scale projects, but indicates the difficulties inherent in the study of lateral movement of invisible points in a soil mass. This technique is known as the use of *horizontal plate gauges.*

The monitoring of lateral movement in smaller structures and smaller earth masses can effectively be carried out by the use of the *inclinometer* system. A cylindrical casing is set up within a specially prepared borehole or fixed to a structure. The casing is originally vertical. After loading or the lapse of time, lateral movement of a structure or a soil mass caused the vertical casing to bend or move and show inclinations to the vertical. It is these inclinations which are measured by the inclinometer. If the inclinations of the casing at various levels are plotted starting at the lowest level of the casing, an outline of the disturbed shape of the casing is prepared. This in turn allows of the lateral movement of any part of the casing to be defined. The *inclinometer,* of which there are several types, consists of a waterproof tube, capable of being lowered down inside the casing. Within the inclinometer tube there is a pendulum whose inclination to the vertical can be reached electrically by monitoring equipment, and whose movement is damped by oil to prevent oscillating readings. The casing usually has some means of controlling the inclinometer tube, so that it follows the deformation of the casing. It should also be noted that the casing must be flexible enough to follow the deformation of the soil in which it is mounted or the wall to which it is fastened. The location of some part of the casing (the top, or perhaps the bottom if that is set in bedrock) must be allowed for, since not only can the casing bend, but it can move bodily from its original position. The original position of the top of the casing, relative to stable points, is known. By surveying from a distance the whole movement of the top of the casing relative to stable ground, can be defined.

Measurement of pressure

The type of pressure most generally measured in confirmatory and research tests is earth pressure. Earth pressure gauges or 'cells' have been the subject of much study with the object of reducing their volume and increasing their accuracy. A metal cell, inserted at the position at which the pressure is to be measured, is an environmental disturbance. Its presence raises doubts as to whether the pressure it records is of the same value as the pressure which would have acted in the absence of the cell. The relatively large movement required of part of the gauge in order to make a recording was at one time a further source of error. However, the movement now required does not invalidate the results, and experience has shown how earth pressure cells may be designed to have the least disturbing effect.

Different means have been used to actuate the measuring device in the cell, but the flexible diaphragm, which is now commonly employed, requires a very small movement and maintains water-tightness in the cell. The diaphragm, when acted on by earth pressure, deflects and so exerts a stress on the *electrical resistance strain gauges* fixed to its inner face. These are sensitive to such stresses. Movement of the diaphragm can be calibrated, by water pressure, against specific stresses. More than one ERSG is required, and four are often used to limit the possibility of the diaphragm recording stresses acting at right angles to those being tested, thus obscuring the true state.

In addition to the electrical resistance strain gauge, the other principal gauge converting pressure to calibrated numerical readings is the *vibrating wire gauge*. A stretched wire, excited into vibration, vibrates at different frequencies according to the stress to which it is submitted. As the diaphragm deflects, the wire is stretched; the frequency of vibration is recorded and transformed to a pressure reading from the results of calibration. The Building Research Station (Thomas and Ward) produced a vibrating wire gauge of a successful type. It shows most of the desirable characteristics required by a pressure cell. Some of these are:

(i) The cell should be as thin as possible in relation to its diameter.

(ii) The stiffness of the diaphragm should be considerably greater than the stiffness of the soil.

(iii) The maximum deformation of the diaphragm should be in the region of 1/5000 part of its diameter.

(iv) The pressure should be applied to the diaphragm over only a small central area $\frac{1}{2}$ to $\frac{1}{3}$ of the area of the diaphragm).

These criteria have been developed on the basis of wide experience, by many workers, of various types of *pressure meter*.

The pressures obtained by these methods are *total pressures*, and by measuring the pore-pressure by the methods given in an earlier section, the *effective pressures* can be obtained.

CHAPTER 14

Laboratory testing and techniques

The object of laboratory testing is to obtain information, additional to that obtainable from *in situ* tests, about the composition and properties of the soil which are essential to the engineer in formulating his designs and in carrying out his works.

Laboratory soil tests may be grouped under the following three headings:

(i) classification and identification tests;
(ii) tests for engineering properties;
(iii) special tests used in engineering construction.

In the first group we have *mechanical analysis* for the determination of the particle-size distribution of the soil. This test is insufficient in the case of fine-grained soils to provide all the information necessary for definite classification, and must therefore be supplemented by the *index tests (liquid and plastic limits)*. In this group should also be placed the tests for the *specific gravity* of the particles and for the *bulk density* and *moisture content* of the soil.

Next come the tests for what may be termed the engineering properties of soils, e.g. *permeability, shear strength, compressibility*.

In the third group are the special tests devised for earthworks and for the subgrades of roads and airfields. Certain of these, such as the California bearing ratio test, are of a comparative nature, no fundamental property being measured directly.

Many of these tests have been standardized, and full instructions are published in B.S. 1377. Procedures for certain tests on stabilized soil are described in B.S. 1924.

14.1 MEASUREMENT OF PARTICLE-SIZE DISTRIBUTION

Mechanical analysis is carried out in two stages:

(i) the separation of the coarser fractions by *sieving* on a series of standard sieves;
(ii) the determination of the proportions of the finer particles by *fine analysis.*

The display of these results is shown in Chapter 2.

Choice of sieves

Before considering the techniques of sieving it is necessary to consider the test sieves themselves. The range of standard sieves in any country is wide, as they are used for many more sieving processes than to obtain soil distributions. The British Standard range consists of 29 large-aperture sizes formed as square or round holes in punched plate, and 36 small-aperture sizes in woven wire cloth. From these, a selection should be made for a specific purpose.

In setting out the particle-size distribution curve, or *grading curve,* logarithmic paper should be used, since it gives the important smaller sizes an adequate separation on the chart. On ordinary squared paper they would be crowded to one end and their significance lost. Logarithmic paper is not always readily available, especially on site, but if the range of aperture sizes is chosen so that any two adjacent sieves in the series maintain a constant ratio of aperture size, ordinary squared paper may be used. The distances between ordinates representing successive sieves are then made equal. This, in terms of logarithmic paper, indicates a constant ratio between successive sieve sizes.

Table 14.1 shows the range of obsolescent sieve sizes which was normally used for sands and gravels. These may still be in use for some time, as they are robust and costly to replace. The ratio between these sieve sizes is approximately two. The British Standard metric aperture

334

sizes are also given for comparison, and one or other of these two ranges should normally be used.

For a quick study of particle-size distribution, the British Standard range of sieves gives the opportunity of studying the particle-size distribution at the sizes which divide soils into specific types

Table 14.1 *Comparison of Imperial and Metric sieves*

Obsolescent sieve sizes which were normally used in particle size analysis	Nearest equivalents in B.S. metric aperture sizes (mm)	
3 in	75·0 S,R	Perforated plate
1½ in	37·5 S,R	
¾ in	20·0 S	
⅜ in	10·0 S	
3/16 in	5·0 S	
No. 7	2·36	Wire mesh
No. 14	1·18	
No. 25	0·60	
No. 52	0·30	
No. 100	0·15	
No. 200	0·075	

S,R–available in square or round holes.
S–available in square holes only.
The ratio of aperture sizes between adjacent sieves in this range is approximately 2.
Equal spacing of successive ordinates gives logarithmic division, accurately enough for assessment of classification.

according to the M.I.T. classification (see Chapter 2). Table 14.2 shows how a shorter range can be used for a quick estimation. This range of sieves has the advantage that from the calculated figures alone, without the need to draw a grading curve, the proportion of the material whose particles are smaller than, say, coarse sand or fine gravel can be read from the computed table of figures. The range of sieves specified in the B.S. 1377, is, however, that shown in Table 14.1.

Table 14.2 *Range of B.S. sieves to match*
B.S. classification

Critical sizes in B.S. classification (see Chapter 2) (mm)	Nearest equivalents in B.S. metric aperture sizes (mm)	
60	63·0 S,R	Perforated plate
20	20·0 S	
6	6·3 S	
2	2·0 R	
0·6	0·6	Wire mesh
0·2	0·212	
0·06	0·063	

S,R—available in square or round holes.
S—available in square holes only.
R—available in round holes only.
The ratio of aperture sizes between adjacent sieves
in this range is approximately 3.
Equal spacing of successive ordinates gives
logarithmic division accurately enough for
assessment of classification.

Methods of sieve analysis
The standard method of obtaining the *particle-size distribution* of a
soil down to the fine sand size is first to wash out the fine material
(silt and clay sizes), and then to dry and sieve the larger sizes. (The
grading of the silt and clay material is then obtained by *fine analysis.*)
The alternative method is to sieve the whole sample of soil in the dry
state, without washing. This latter method is very often adopted
because of the irksome procedure required for wet sieving. *Dry
sieving,* however, is not accurate if there is an appreciable quantity of
fine material in the sample.

Wet sieving. The specification for this test (given in B.S. 1377) is
directed to the separation of all silt and clay from the sample without
the loss of any fine sand. The use of dispersal agents and various
controlled washings makes sure that the material retained contains all
the particles with sizes greater than 75 μm (0·075 mm). The fine
material, less than this size, is washed to waste and the larger, retained
material dried and sieved.

Dry sieving. In this process the whole sample is dried and passed through the sieves without any of the fine material being first extracted. This is a faster method, but not so accurate if there is a substantial proportion of fines. The precautions to be taken in both wet and dry sieving are listed in B.S. 1377.

In either method, dry material is passed down through a nest of sieves from the larger to the smaller sizes. The material retained on each sieve is weighed, and tables such as those shown in the Applications are prepared.

Fine analysis

The process known as *fine analysis* is based on Stokes' law, according to which fine particles from suspension in a liquid settle at different rates according to their size, the coarser settling more quickly than the finer. Stokes' law is true for spherical particles only, and since soil grains are not spherical in shape is it usual to define *particle-size* in terms of the *effective diameter,* that is the diameter of an imaginary sphere of the same material which would sink in water with the same velocity as the irregular particle in question. Particles less than about 0·0002 mm effective diameter do not settle in accordance with Stokes' law, and the sedimentation method of mechanical analysis does not, therefore, yield any information about the proportion of particles smaller than this.

The velocity of a particle, sinking in a still fluid, is given by Stoke's law as

$$v = \frac{2}{9} g r^2 \frac{\gamma_s - \gamma_w}{\eta}$$

when the symbols (see Notation) are evaluated in consistent units. Using millimetres and seconds, and assuming that the fluid is water with a density of 10^{-3} g mm^{-3}, and that the solid particles have a specific gravity of 2·65, this expression reduces to

$$v = 0·9 D^2 / \eta \text{ mm s}^{-1}.$$

The dynamic viscosity of water is 0·001 g mm^{-1} s^{-1} at 20°C. There is a small variation with temperature, but if the basic value is taken, then

$$v = 900 D^2 \text{ mm s}^{-1}.$$

The time t for a particle of effective diameter D to sink through

100 m is, thus

$$t = 9/D^2 \text{ s}$$

and

$$D = \sqrt{\left(\frac{h\eta}{0 \cdot 9t}\right)} \text{ mm.}$$

Table 14.3 shows the times required for various particle sizes.

Table 14.3 *Sedimentation times (for* $\eta = 0 \cdot 001 \ g \ mm^{-1} s^{-1};$ $G_s = 2 \cdot 65;$ *settlement through 100 mm)*

D (mm)	t (s)
0·06	31
0·02	278
0·006	3 086
0·002	27 778
0·001	111 111

A suspension of soil particles contains an assortment of grains of various sizes, and the distribution is assumed to be uniform at the beginning of the test. After time t all particles of diameter D and over will have settled through a depth h. At this depth, therefore, all the particles left in suspension will be of diameter less than D, and these particles will be in the same concentration as at the commencement of the test.

Thus the percentage of particles less than D is equal to

$$\frac{\text{Weight of solids per ml at depth } h \text{ after time } t}{\text{Weights of solids per ml in original suspension}} \times 100 = \frac{W_D}{W} \times 100.$$

The detailed procedure for this and other tests in this chapter are given in B.S. 1377, and in equivalent national codes and specifications, but a brief description is given in this chapter.

Standard and subsidiary methods of fine analysis
It should be remembered that these methods have, as their object, the grading of material of the silt and clay sizes with some of the finer

338

sand included if necessary. If the material shows less than 10 per cent of its weight passing the 200 mesh sieve, it is unsuitable for this type of analysis, and should be tested by wet or dry sieving.

In both the standard and subsidiary methods of fine analysis, the suspension of the fine particles is ensured by the use of a dispersal agent. The suspension is brought to the standard temperature (25° C) in a water bath, and then the cylinder containing the liquid is thoroughly shaken and set vertically in the water bath. A stop watch is started (to record the times of sedimentation) as the tube is set in its upright position.

Standard method. Three samples are drawn off by a special pipette at specified time intervals which relate to the diameter of the particle and to the specific gravity of the particles being tested. The times given in Table 14.3 are examples of those theoretically obtained for one specific gravity, and one assumed value of dynamic viscosity. Those specified in B.S. 1377 relate to 0·02, 0·006 and 0·002 mm. Each of the samples (taken at a depth of 100 mm below the surface) contains only particles of a smaller size than the one quoted for the test. Thus, the 'percentage-passing' can be calculated and a grading curve drawn for the finer particles.

Subsidiary method. Here, the specific gravity of the suspension is measured by means of a hydrometer. Readings are taken at $\frac{1}{2}$, 1, 2, 4, 8, 15, 30 min and at 1, 2 and 4 h after the start of sedimentation. Further readings over a period of a few days can also be made. The relating of the values of the specific gravity of the suspension to the values of the particle sizes remaining in the suspension at that time, is usually carried out by a nomogram based on Stokes' Law.

In employing both of these methods, considerable care must be exercised. The dispersion of the particles so that they do not 'clump' together, the calibration of the apparatus, the maintenance of the correct temperature, and the application of various corrections make these tests somewhat more demanding than the carrying out of a sieving operation for the larger particle sizes.

14.2 SPECIFIC GRAVITY AND MOISTURE CONTENT

The terms *bulk density, dry density, moisture content,* and *void ratio* are fully discussed in Chapter 2. In this section various tests are described from which these important properties can be determined.

Specific gravity of soil particles

For fine-grained soils, the specific gravity may be determined by the use of a 50 ml density bottle by the methods employed in physics laboratories. For coarse-grained soils much larger samples are required to give reliable results. A *gas cylinder* of one litre capacity replaces the small density bottle (Fig. 14.1). A sample which can lie between 200 g for finer soils and 400 g for coarser soils is over-dried and put into the gas jar which is then weighed, together with its ground-glass cover.

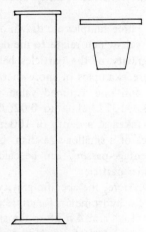

Fig. 14.1. Proportions of litre jar, rubber bung, ground glass plate for specific gravity test.

Water is then added (about 500 ml) and by techniques described in B.S. 1377, the air is removed and more water then added until the jar is full. The ground glass cover is slid into position over the water.

At one stage in the preparation, the gas jar must be turned over end to end in a shaking apparatus for up to $\frac{1}{2}$ h. To effect this without loss of contents, the jar must be closed by the rubber bung which forms part of the equipment for the test.

Four masses are recorded by weighing

Jar and ground glass cover (empty)	(m_1)
Jar and ground glass cover and dry soil	(m_2)
Jar and ground glass cover, soil and water	(m_3)
Jar and ground glass cover and water to fill jar.	(m_4)

The mass of the soil is given by $(m_2 - m_1)\,g$.

The mass of water filling the jar is $(m_4 - m_1)\,g$. This is also the volume of the jar in ml.

The mass of water in the jar, when the soil is occupying some of the volume, is $(m_3 - m_2)\,g$. This is also a volume in ml.

The volume occupied by the soil particles, all air having been extracted is $(m_4 - m_1) - (m_3 - m_2)\,ml$.

Thus the specific gravity of the soil particles $(g\,ml^{-1})$ is

$$\frac{\text{Mass of soil}}{\text{Volume of soil}} \text{ or } \frac{(m_2 - m_1)}{(m_4 - m_1) - (m_3 - m_2)}$$

To obtain accurate results it is essential that the techniques described in B.S. 1377 should be followed closely.

Determination of moisture content

In the standard method the soil is placed in a glass weighing bottle or non-corrodible container. The bottle or container with contents (whose mass is known) is then placed in an oven maintained at 105-110° C until no more weight is lost. The difference between the original and final weighings represents the mass of water evaporated.

m_1 is mass of container (g)

m_2 is mass of container and wet soil (g)

m_3 is mass of container and dry soil (g).

$$\text{Moisture content, } w = \frac{m_2 - m_3}{m_3 - m_1} \times 100\%.$$

For fine-grained soils the container may be as small as 50 mm diameter and 25 mm high. For medium-grained soils a sample of 500 g must be contained and dried and for coarse-grained soils 3 kg must be contained and dried.

B.S. 1377 gives alternative methods of moisture content determination which are quicker, though less accurate. In one of these the soil is dried over a heated sand-bath; in the other the water is evaporated by burning methylated spirit in contact with the soil.

14.3 CONSISTENCY LIMITS

The determination of the consistency limits forms an important part of the classification tests. In particular, the liquid and plastic limits and the plasticity index of cohesive soils give a useful indication of their general characteristics.

As explained in Chapter 2, the liquid, plastic, and shrinkage limits mark the transitions between the liquid, plastic, semi-solid, and solid conditions respectively. As these changes of state are not sharply defined it is necessary to use arbitrary definitions and to specify standard tests.

Determination of liquid limit

There are two acceptable tests to determine the *liquid limit* of a cohesive soil. The second of these tests, by *Casagrande's apparatus*, has been used for many years, but close agreement of repeated results depends on acquired skill and judgement to a greater degree than by the *Core-penetrometer method* which is preferred.

By cone-penetrometer. The cone is of stainless steel or duralumin with an angle of 30° and a length of 35 mm. It is important that the point of the cone remains sharp, and details of the precautions to be observed are given in B.S. 1377.

The cone is allowed to sink into a standard sample of wet soil, the mass of the cone and its supporting rod being 80 g.

Several tests are carried out at increasing moisture contents and penetrations achieved are plotted against moisture content. From this graph, the moisture content at which there is 20 mm of penetration is recorded as the *liquid limit*.

This test may also be carried out on soil in its natural state, but the results will usually differ from those obtained from prepared samples. It is important to record that the figure obtained is obtained by the cone-penetration method, and also whether the more rigorous method or soil in its natural state has been used.

The *liquid limit apparatus* designed by Casegrande is shown in Fig. 14.2. A cake of wet soil is placed in the circular brass dish, and a groove cut in it with the grooving tool. By means of a handle rotating a cam the dish is raised to a height of 1 cm and then falls freely on to a hard block of rubber. The soil is said to be at the *liquid limit* when 25 blows are required to close the gap. It is essential that a new liquid limit apparatus should be calibrated against a standard machine, as the hardness of the block is a critical factor.

The method of procedure is to mix a sample of dried and powdered soil with water to a stiff consistency. The cam is rotated at

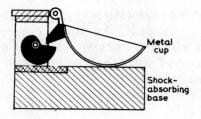

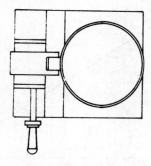

Fig. 14.2. Liquid limit apparatus

the rate of about two blows per second until the groove closes over a length of about 13 mm, and the number of blows to effect this is recorded. A small portion of the soil from the dish is then put into a container for moisture content determination. The test is repeated after successive additions of water. The results of the tests are recorded as a 'flow curve' (Fig. 14.3) in which the moisture content values are plotted to a natural scale against the number of blows to a logarithmic scale. A straight line results, from which the moisture content for 25 blows is obtained by interpolation. It is desirable to obtain two or three experimental points on each side of the liquid limit.

The one-point method. A faster and less troublesome method of obtaining the liquid limit is to use only one experimental determination. Some experience of the liquid limit test is advantageous. The

343

operator, in forming the soil/water mixture, should try to estimate, by the appearance of the paste, that the mix approximates to the moisture content relating to 25 blows in the standard method. Only one such mix is made. The paste is mixed for at least 10 min and then stored in an air-tight container for 24 h.

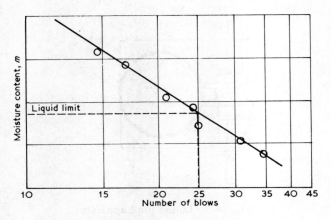

Fig. 14.3. Assessment of liquid limit

The next day the mixture is tested in the *liquid limit apparatus* until two consecutive tests give the same number of blows. If this number of blows does not fall between 15 and 35, the mix must be remade and the test started again with a better estimate of the consistency of the paste. When the test is completed satisfactorily, the moisture content is then obtained and multiplied by a factor according to the number of blows required in the one-point test. The resulting figure is the liquid limit. The factors are given in B.S. 1377 and vary from 0·95 for 15 blows, to 1·03 for 35 blows.

Determination of plastic limit
A wetted sample of about 20 g (passing the 425 μm sieve) is rolled with the palm of the hand on a glass plate into a thread of about 6 mm diameter. It is folded and rolled repeatedly until the 6 mm

344

thread begins to crumble. The moisture content of the crumbled sample is then determined, and this is the *plastic limit*.

The liquid and plastic limits having been determined, the *plasticity index* is found by subtraction. The use of the values of liquid limit and plasticity index to determine the classification and properties of cohesive soils is discussed in Chapter 2.

Caution
The descriptions of tests given in this CHapter are intended as introductions to techniques and as a quick survey of requirements. Exact and detailed procedures are required to obtain reliable results, and original specifications must be consulted.

14.4 MEASUREMENT OF PERMEABILITY

The permeability of coarse-grained soils is easily determined in the laboratory by means of the *permeameter*. The principle of the test is to cause water to flow through a sample of soil, and to measure the hydraulic gradient and the rate of discharge. Then, by applying Darcy's equation, the coefficient of permeability can be calculated. As the permeability of a given soil varies with its density, it is important to pack or tamp the soil in the permeameter to the density at which the coefficient of permeability is required. Two forms of permeameter are commonly used which are known respectively as the constant head permeameter and the falling head permeameter.

For fine-grained soils of low permeability (see Fig. 4.1) the permeameter method is impracticable, and the coefficient of permeability can be determined only by indirect methods. The most important of these is the calculation of the coefficient from the results of the consolidation test.

Constant head permeameter
The sample of soil is placed in a cylinder of cross-sectional area A, and water is allowed to percolate through under a constant head (Fig. 14.4). The discharge Q during a suitable time interval t is collected in a graduated vessel. By means of tappings in the cylinder connected to manometers the difference of head H over a length of

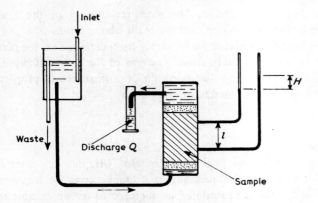

Fig. 14.4. Constant-head permeameter (diagrammatic)

sample *l* is measured. Then by Darcy's law

$$\frac{Q}{t} = Aki = Ak\frac{H}{l}$$

from which *k* is calculated

Before starting the test, the apparatus is subjected, through a connection not shown in the diagram, to the action of a vacuum pump in order to extract all air from the pores of the sample.

Falling head permeameter

For very fine sands and silts the falling head permeameter (Fig. 14.5) is more suitable. In this device the cylinder containing the sample is fitted with a cap and standpipe. The apparatus is filled with water up to a mark on the standpipe. The stopcock is opened and the water allowed to percolate through into the container at the base until the level in the standpipe has dropped to a second mark. The time *t* taken for the water level to fall this distance is observed.

Let H_1 and H_2 be the initial and final heights of the water in the standpipe above that in the container, and let *H* be the height at an intermediate time. Let *A* be the cross-sectional area of the sample and *a* that of the standpipe. Let *l* be the length of the sample.

During a time interval *dt* the water-level drops *dH*, i.e. the increment of *H* is −*dH*. By Darcy's law the discharge in time *dt* is

$$Ak\frac{H}{l}\,dt.$$

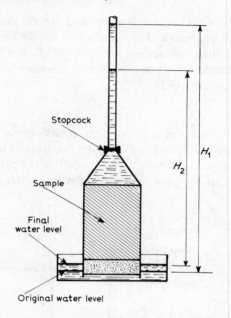

Fig. 14.5. Falling-head permeameter (diagrammatic)

Therefore

$$-adH = Ak\frac{H}{l}\,dt$$

from which

$$dt = -\frac{a}{A}\frac{l}{k}\frac{dH}{H}.$$

Integrating both sides between corresponding limits,

$$t = -\frac{a}{A}\frac{l}{k}\log_e\frac{H_2}{H_1} = \frac{a}{A}\frac{l}{k}\log_e\frac{H_1}{H_2}.$$

The quantity $(a/A)l\,\log_t(H_1/H_2)$ is a constant for the particular permeameter.

14.5 MEASUREMENT OF SOIL SUCTION

The importance of the property known as soil suction is discussed in Chapter 3. Several methods have been devised for measuring this property. Two of the most frequently used are the suction plate and the membrane apparatus. The former can be used only when the suction is less than atmospheric (*pF* 0 to *pF* 3), but the membrane method is suitable for a range of suction pressures from 0·1 to 100 atmospheres (*pF* 2 to *pF* 5).

Suction plate
A form of this apparatus is shown diagrammatically in Fig. 14.6. The soil sample rests on a sintered glass plate, the pores of which are small enough to prevent air from entering, yet allow the passage of water. The underside of the plate is in contact with a water-filled reservoir

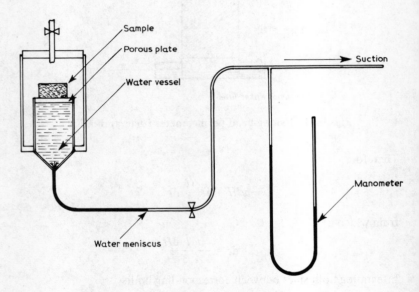

Fig. 14.6. Suction-plate apparatus (diagrammatic)

connected to a horizontal glass tube. The tube is connected to a vacuum pump, with a mercury manometer to measure the suction.

When a sample of soil is placed on the suction plate, water tends to be drawn into it and the water meniscus in the tube travels towards the reservoir. When sufficient vacuum is applied to prevent movement of the meniscus, the suction indicated by the manometer is equal to the soil suction corresponding to its moisture content. To obtain the soil suction/moisture content relationship, several samples are tested at different amounts of vacuum. The moisture content of each sample is measured after water has been drawn in or out until equilibrium is reached.

Membrane method

This apparatus, shown diagrammatically in Fig. 14.7, consists of a pressure cell of robust construction in which the soil sample rests on a porous bronze disc, separated by a thin cellulose membrane. The membrane has the property of being impermeable to air but permeable to water. After the sample has been placed in position, the

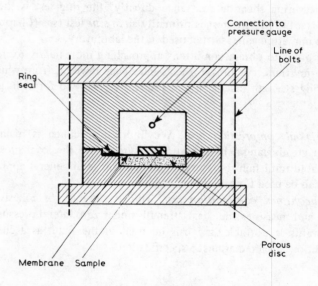

Fig. 14.7. Membrane apparatus (diagrammatic)

cell is bolted together and compressed air is admitted at the required pressure. The pressure causes water to be expelled from the soil through the membrane until equilibrium is reached, when the applied pressure just balances the soil suction. The pressure is usually maintained for 24 h, after which it is released. The cell is then dismantled and the final moisture content of the soil measured.

14.6 MEASUREMENT OF SHEARING RESISTANCE

Tests for the shearing resistance of soils may be divided into:

> *direct shear tests:* shear box, vane and ring shear.
> *indirect shear tests:* triaxial compression and unconfined compression.

The original form of apparatus for the direct application of shear force is the *shear box*. This test, though simple in principle, is open to the objection that the stress distribution across the specimen is complex, and the value of the shearing resistance obtained by dividing the shearing force by the area is only approximate. Of other methods of measuring shearing resistance directly, the ring test is the most important. The vane test is primarily an *in situ* test (see (Chapter 13), but a miniature vane is often used in the laboratory.

In order to obtain conditions approaching more nearly to those of uniform stress, compression tests have been devised, from which the shearing strength can be found indirectly. The two forms in general use are:

> *Triaxial compression test.* A cylindrical specimen is maintained under steady lateral hydrostatic pressure, while the axial pressure is increased until failure occurs. The apparatus is of general application and can be used for either sand or clay.
> *Unconfined compression test.* This is suitable for cohesive soils only and measures the shear strength under zero lateral pressure. The apparatus is portable and can be used in the field, as a check on conditions attained against a specification.

Shear box
The apparatus (Fig. 14.8) consists of a square brass box split horizontally at the level of the centre of the soil sample,which is held between metal grilles and porous stones. A gradually increasing horizontal load is applied to the upper part of the box until the sample fails in shear. The shear load at failure is divided by the cross-sectional area of the sample to give the ultimate shearing stress. When the shearing resistance under a normal pressure is required, a vertical load is applied to the sample through the piston by means of dead weights.

In order to apply the load at a constant rate of strain, the lower half of the box is mounted on rollers and is pushed forward at a uniform rate by gearing. The upper half of the box bears against a steel proving ring, the deformation of which is shown on the dial gauge and indicates the shearing force. To measure the volume change during consolidation and during the shearing process another dial gauge is mounted to show the vertical movement of the piston. Water escapes or enters the specimen through the porous stones and the perforated metal grilles.

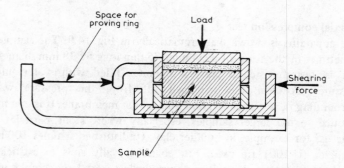

Fig. 14.8. Constant-rate-of-strain shear box

The size of shear box normally used for clays and sands is 60 mm square and the sample is 20 mm thick. A larger size, 300 mm square, is frequently used for gravelly soils, where the sizes of the stones demands a larger system.

Residual shearing strength

The measurement of residual shearing strength can be made by using the triaxial test, but, since large strains are required to reach the residual strength, it is more than likely that sufficient movement cannot be obtained to reach the residual stage. Ring shear, in which the sample is subjected to large torsional strains, has the advantage of providing continuous and uniform deformation through whatever strain is required. The shear box is the third and probably the most frequently used method. The apparatus is adapted and can be automatically controlled to provide reversing direct shear. As the carriage of the box travels in both directions, the proving ring, for example, must be connected to read tensile as well as compressive loads as the direction of shear is reversed. The importance of knowing the value of the residual shear strength in certain circumstances, particularly with over-consolidated clays, is described in Section 5.4.

Laboratory vane test
The vane test is primarily an *in situ* test (Chapter 13), but a smaller size of vane is used in the laboratory for soft clays. The method of calculating shearing strength from the measured torque applied to the vane is described in Section 13.2.

Triaxial compression test
The apparatus is shown diagrammatically in Fig. 14.9. The sample is sometimes in the form of a cylinder 75 mm long by 38 mm diameter, and is tested inside a perspex cylinder filled with water under pressure. To seal the specimen off from the pressure water surrounding it, it is enclosed in a thin rubber membrane. It is common practice to use larger samples, especially where stone particles are large as, for example, in boulder clay. Undisturbed samples 100 mm diameter (U 100) are used. For coarse gravelly soils or artificially prepared granular soils such as railway ballast, very large samples with appropriately large testing equipment are required if realistic values of shearing strength are to be obtained.

The pressure is raised to the desired value by means of a pump, and vertical loading is applied by moving the piston rod downwards at a constant rate. The load is measured by a steel proving ring, the deformation of which is shown on a dial gauge.

For clay, undisturbed samples are trimmed to the correct size. For consolidated tests the metal caps and bases are provided with porous plates to enable drainage to take place during consolidation. During the compression test the outlet valve is shut for undrained conditions and open for drained. For undrained tests solid end pieces are substituted. When pore-pressure measurements are to be made porous ends are used and the outlet is connected to the pore-pressure measuring apparatus.

The use of the rubber tube enables tests to be made on non-cohesive soils. Loose material is filled into the rubber tube, temporarily supported in a metal former. The sample is then placed in the machine and flooded with water through the porous base. A vacuum is applied to ensure that all air is extracted, and the metal

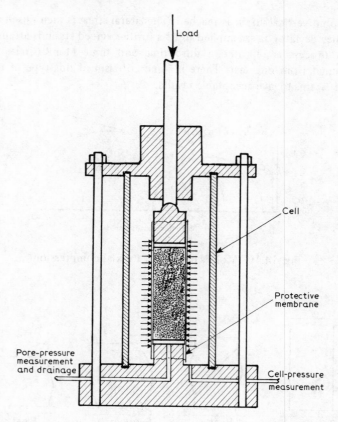

Fig. 14.9. Triaxial-compression apparatus (diagrammatic)

support is removed. The perspex cylinder is then placed in position and the lateral pressure applied.

It should be noted that the load shown on the proving ring gauge indicates the *deviator* stress or principal stress difference, $\sigma_1 - \sigma_3$. Two tests at different lateral pressures are necessary to determine c and ϕ, but in practice three or more tests are usually carried out. For each test the total axial pressure at failure σ_1, and the lateral pressure σ_3 are plotted as shown in Fig. 14.10 and the Mohr Circle drawn.

Multi-stage triaxial compression tests are carried out on 100 mm diameter samples. With a low lateral pressure (to provide the smallest Mohr Circle, Fig. 14.10) the deviator stress is applied until an

353

accepted vertical strain is reached. The lateral stress is then raised and further deviator stress applied until a further agreed strain is attained. This process is repeated a third time and three Mohr Circles are obtained from one test. There is some criticism of this type of test, but it seems to give acceptable results.

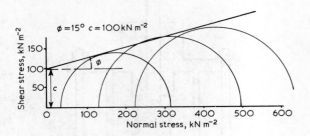

Fig. 14.10. Mohr's circles for triaxial compression

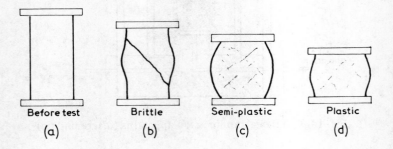

Fig. 14.11. Types of compression failure

The failure of compression specimens usually takes one of the forms shown in Fig. 14.11. In the brittle type of failure the specimen shears on diagonal planes. In the plastic type the material bulges and ultimately fails by sliding simultaneously on numerous planes. The former type of failure is generally obtained with sands and sandy clays, and sometimes with stiff clays. Soft plastic clays usually fail by bulging, and this type of failure is sometimes obtained from the triaxial test on certain sands with a high void ratio.

Unconfined compression test

The apparatus for this test is portable and can therefore be used in the field. Undisturbed samples are used, of the same dimensions as for the triaxial test, 76 mm long by 38 mm in diameter.

The specimen is loaded in compression until failure takes place, either by shearing or a diagonal plane or by lateral bulging. The compressive stress (load per unit area) is then the greatest principal stress, σ_1, and the other two principal stresses are zero. Since the angle of shearing resistance is assumed to be zero, the shear strength is taken as half the ultimate compressive strength. The test is used for undrained tests on cohesive soils.

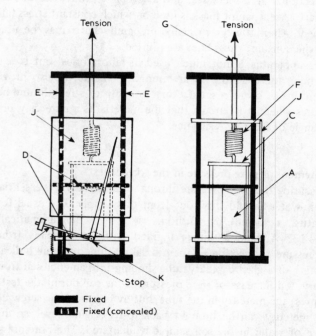

Fig. 14.12. Unconfined-compression apparatus

The machine is shown in Fig. 14.12. The specimen A is placed between two metal cones carried on horizontal loading plates D. The upper plate is fixed, while the lower one slides on vertical rods. The plate C is supported through the spring F by a screw G and can be raised by turning a handle so as to apply a compressive load to the specimen.

The stress-strain diagram is plotted autographically. The vertical chart plate *J* moves upwards with the yoke to which the top of the spring is attached. The pen arm pivot *K* moves upwards with the lower end of the specimen; thus the vertical movement of the pen relative to the chart is equal to the extension of the spring and therefore proportional to the load. As the lower plate *D* moves upwards the pen swings sideways, the weighted arm *L* bearing on a stop. The lateral movement in arc of the pen is thus proportional to the compressive strain on the specimen. As the specimen is compressed its cross-sectional area increases, and in order to read stress direct from the chart a transparent mask is used on which constant stress lines are inscribed. Alternatively, specially prepared charts may be used, on which the constant stress lines are printed.

This apparatus is not much used in laboratories, but is a useful instrument for checking shear strengths on site. It also provides a rapid means of giving a satisfactory estimate of strengths and bearing capacities where it is known that the material is a pure clay or has a small angle of shearing resistance.

Measurement of pore-pressure in the triaxial test

In measuring the pore-pressure during a triaxial test it is essential that no pore-water should flow out from the sample. The usual form of apparatus, designed by Bishop, is shown diagrammatically in Fig. 14.13. A flexible tube *A* filled with water leads from the pore-pressure connection of the triaxial machine to a null indicator *B*, consisting of a glass capillary tube dipping into an enclosed trough of mercury. An increase of pore-pressure in the soil during the test tends to depress the mercury in the tube, but by means of the screw control *C* the mercury can be brought back to its original level, so that the volume of water in the soil sample is unaltered. The pore-pressure is shown either on the pressure gauge *D* or on the mercury manometer *E*. The latter is used for measuring negative pore pressures and for the more accurate measurement of small positive pressures. The burette *F* is used for determining the pressure gauge and manometer readings for zero pore-pressure. It can also be used for the measurement of volume change during triaxial tests in which drainage from the specimen is permitted.

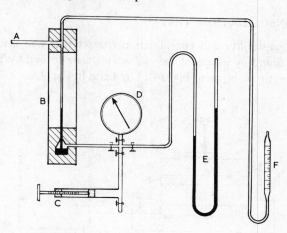

Fig. 14.13. Measurement of pore-pressure in the triaxial-compression test

Great care has to be taken that the parts of the apparatus containing water are completely free from air, as even minute air bubbles are liable to cause serious errors in the readings.

Measurement of pore-pressure coefficients

As explained in Chapter 3 the pore-pressure coefficients are related by the equation

$$\Delta u = B[\Delta\sigma_3 + A(\Delta\sigma_1 - \Delta\sigma_3)] = \bar{B}\Delta\sigma_1.$$

Measurement of B. For the condition of equal stress all round, $\Delta\sigma_1 - \Delta\sigma_3 = 0$ and the A-term in the above equation disappears. Hence B can be determined by simple pore-pressure measurement in the triaxial test with equal all-round stresses.

Measurement of $\bar{B}$ and A. The combined coefficient $\bar{B}$ is first found from a controlled stress–ratio test, that is one in which the principal stresses σ_1 and σ_3 are increased proportionally. For each value of the pressures the pore-pressure u is measured and $\bar{B}$ is found from the equation

$$\Delta u = \bar{B}\Delta\sigma_1.$$

The, since B and $\bar{B}$ are known, A can be calculated.

357

14.7 CONSOLIDATION TEST

The compressibility and consolidation characteristics of fine-grained soils are found by means of the oedometer test, devised by Terzaghi. The apparatus is shown in Fig. 14.15 and also in Fig. 9.6.

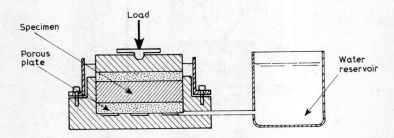

Fig. 14.14. Oedometer or consolidation apparatus (diagrammatic)

The specimen of soil is contained in a flat cylindrical mould between two porous discs. The diameter of the mould should be slightly smaller than that of the undisturbed sample from which the specimen is to be cut. The depth of the mould should be one third to one quarter of the diameter. A common size is 75 mm diameter by 19 mm deep. By means of the porous discs water has free access to and from both surfaces of the specimen. The compressive load is applied to the specimen through a piston, either by means of a hanger and dead weights or by a system of levers. The compression is measured on a dial gauge.

In the test for compressibility, a load is applied to the specimen and readings of the compression are taken at appropriate time intervals after the application of the load. The consolidation is rapid at first, but the rate gradually decreases. After a time the dial reading becomes practically steady, and the sample may be assumed to have reached a condition of equilibrium. For the size of specimen used in the test, this condition is generally attained within 24 h, though theoretically an infinite time is required for complete consolidation. The method of working out the results of this test is described in a previous chapter.

Field measurements of drainage have shown that the 75 mm diameter oedometer test may give misleading results unless the clay is uniform. For non-homogeneous clays specimens 250 mm diameter by 125 mm thick appear to represent more nearly the natural structure of the soil.

14.8 SOIL TESTS FOR ENGINEERING CONSTRUCTION

The tests in this group include the standard tests for compaction, the California bearing ratio test for the subgrades of roads and airfields and various tests for the properties of stabilized soil.

Standard compaction test   (see Section 11.2)

In this test, often known as the Proctor test, a cylindrical mould 950 mm^3 in volume is filled in three layers, each layer of the soil being compacted by 25 blows of a standard rammer which has a mass of 2·5 kg and drops 300 mm at each blow. The mould is fitted with a detachable collar which enables the soil to be compacted to a level slightly above the rim of the mould itself. The collar is then removed and the surface carefully struck off level with the rim. The bulk density is obtained by weighing the mould, and a small portion of the soil is taken for moisture content determination. From these measurements the dry density is calculated. The test is repeated for a wide range of moisture contents, and a curve is plotted showing dry density on a base of moisture content (see Fig. 11.1). From this the maximum dry density and the optimum moisture content are read off.

In the heavy compaction test, based originally on the American Association of State Highway Officials' standard, the same mould is used by the compaction is effected by 27 blows of a 4·5 kg rammer dropped through 450 mm on each of five layers of soil.

As an alternative method of compaction, a vibrating rammer of a mass of 3 kg may be used with a mould 152 mm diameter and 127 mm high (this is the California Bearing Ratio mould described in the next section). The vibrating rammer method is suitable for granular soil and has been found to give results which are closer to field conditions than are those from the normal methods.

California bearing ratio test

This test, which is carried out on material passing the 20 mm test sieve, is used in the determination of the thicknesses of pavement required for roads and runways. It shows its best results when the soils tested are not wet, weak and cohesive, but show some resistance to penetration. The test is complex; the full procedure is described in specifications of various countries, as it is internationally accepted. The care with which the sample of soil is prepared for testing has a marked influence on the value of the results obtained; it is essential to follow the specification closely (nine pages long in B.S. 1377).

In brief, the preparation consists of filling the soil into a cylindrical mould 152 mm diameter and 127 mm high by specific techniques required by the specification. The mass of soil filling the mould is constant and exactly weighed. The soil is compressed into the mould so that the mould is filled to the upper rim. This is done either by static compression in a testing machine or by dynamic compression by the use of the rammers of the standard compaction test. A collar is fitted above the mould during the filling in order that there is no loss of material.

At this point the sample may be soaked by allowing water to percolate up through a perforated base plate. The soaking is intended to simulate the worst condition likely to be attained by the soil as part of a road foundation. For roads and runways with impervious surfaces and adequate drainage this technique may result in unnecessarily conservative design. It is unusual for the soaking procedure to be used in the U.K.

The Test. A standard circular plunger (1935 mm^2 in cross-section) is seated on the top of the specimen with a load of 50 or 250 kN, depending on the likely resistance of the soil. It is then driven into the soil and readings of the force required made at every 0·25 mm. The total penetration does not exceed 7·5 mm.

The graph of force against penetration is then drawn and compared with the graph for a standard material. At 2·5 mm and again at 5 mm penetration, the ratio

$$\frac{\text{Force required on tested soil}}{\text{Force required on standard soil}}$$

is read off the two graphs, and the larger of these two ratios is recorded as the C.B.R. The shape of the load/penetration curve for the tested soil is important. It is usually concave downwards at the lower end, but if concave upward, a correction must be made or quite misleading results will be obtained. The forces on the plunger are plotted as ordinates and the penetrations as abscissae.

In cohesive soil the test is sometimes made on an undisturbed core. For this purpose a cutting edge is fitted to the CBR mould and this is pressed into the ground as in ordinary undisturbed sampling. The CBR test is also sometimes carried out *in situ*; here the absence of the restraint imposed by the mould makes the conditions of the test slightly different.

This test is also used as a comparative test of the quality of stabilized soil, and the procedure for this purpose is described in B.S. 1924. When used in this way the result obtained from the test is known as the *cylinder penetration ratio* (CPR).

Tests for stabilized soil

Several simple tests have been devised for measuring the properties of stabilized soil and for determining the most suitable proportions of stabilizing material. These are fully specified in B.S. 1924.

Compression test. The dry soil, with a definite proportion of stabilizer, is thoroughly mixed with a known quantity of water. The optimum moisture content as determined by the standard compaction test is a useful guide as to the quantity of water to be used. For fine-grained soils the test specimens are of cylindrical form, 100 mm high by 50 mm in diameter. Larger specimens of the same proportions are used for medium- and coarse-grained soils. The moulds are filled either by static compaction, in which the density if controlled, or by dynamic compaction, in which the controlling factor is the compactive effort. The specimens are extruded from the moulds, coated with paraffin wax and then cured in air at a uniform temperature. After seven days, or other specified period, the paraffin wax is carefully removed from the ends of the specimens and they are tested in compression.

By carrying out a series of tests with different proportions of stabilizing agent it is possible to decide on the best proportion to be used in the field.

Immersion test. The object of this test is to compare the compressive strength of two specimens, one of which has been cured in the normal way for fourteen days, while the other has been cured for seven days and then immersed in water for seven days. The paraffin wax is carefully removed from the latter specimen before immersion. The resistance to softening is found by expressing the compressive strength of the immersed specimen as a percentage of that of the control specimen.

Resistance to frost damage. In this test the compressive strength of a specimen which has been subjected to a number of specified cycles of immersion, freezing, and thawing is compared with that of a normally cured specimen.

Water absorption test. Specimens are prepared as for the unconfined compression test. After normal curing for three days the wax is removed and the specimens are immersed in water for various periods from 1 to 28 days. The increase in weight due to absorption of water is recorded.

Other tests

Tests for the proportion of organic matter in the soil, for the sulphate content of both water and soil samples, and for acidity (pH) are also described fully in B.S. 1377.

Reference list

The following books, pamphlet and technical papers form a selection which will give the student a wider insight into the theories and applications of the principles of soil mechanics. The list has been deliberately restricted so that the importance of the references given will be all the more apparent. There are many more which could be consulted. The Codes of Practice and Standards are directed to readers in the United Kingdom, but similar Codes and Standards are published in other countries. Reference should be made to those which are appropriate for the working conditions. The latest editions (of the works quoted in the list) should be consulted.

Textbooks
Terzaghi, K., *Theoretical Soil Mechanics,* John Wiley and Sons, London and New York.
Terzaghi, K. and Peck, R. T., *Soil Mechanics in Engineering Practice,* John Wiley and Sons, London and New York.

Testing
British Standard 1377: Methods of testing soils for civil engineering purposes.
British Standard 1924: Methods of test for stabilized soil.
Bishop, A. W. and Henkel, D. J., *The Measurement of Soil Properties in the Triaxial Test.*

Pore-pressure, foundation pressure, settlement
Skempton, A. W. (1954). The pore-pressure coefficients A and B, *Geotechnique* **4**, 143.
Bishop, A. W. (1954). The use of pore-pressure coefficients in practice, *ibid* 148.

Newmark, N. M. (1942). Influence charts for computation of stresses in elastic foundations, *Univ. of Illinois Bulletin* No. 338.

Skempton, A. W. and MacDonald, D. H. (1956). The allowable settlement of buildings, *Proc. I.C.E.* **5**, Part III, 727.

Codes of Practice

British C.P. 101: Foundations and sub-structures for non-industrial buildings of not more than four storeys.

British C.P. 102: Protection of buildings against water from the ground.

British C.P. 2001: Site investigation.

British C.P. 2003: Earthworks.

British C.P. 2004: Foundations.

Index